BARRON'S
NEW YORK STATE

GRADE 3 MATH TEST

SECOND EDITION

Margery Masters, M.S. Ed.

T0116645

About the Author

Margery Masters is an elementary school math specialist and the author of several elementary-level math books for Barron's.

Dedication

This book is dedicated to my parents, Arnold and Ona Masters, with thanks for their love and support, and to Mattye Streit, for her loving care of my parents.

Acknowledgments

I would like to thank the following for their contributions to this book: Brenda Clarke, Ona Masters, and Susan Yoder for their proofreading assistance; and third grade teachers Jan Cosgrove, Bethany Deyermond, Kurt Kahofer, and Maria Semkus for their knowledge and support.

© Copyright 2012, 2008 by Barron's Educational Series, Inc.

All rights reserved.
No part of this publication may be reproduced or distributed in any form or by any means without the written permission of the copyright owner.

All inquiries should be addressed to:
Barron's Educational Series, Inc.
250 Wireless Boulevard
Hauppauge, NY 11788
www.barronseduc.com

ISBN: 978-1-4380-0042-8
ISSN: 1942-0137

Date of Manufacture: January 2017
Manufactured by: B11R11, Robbinsville, NJ

Printed in the United States of America
9 8 7 6 5

10%
POST-CONSUMER
WASTE
Paper contains a minimum
of 10% post-consumer
waste (PCW). Paper used
in this book was derived
from certified, sustainable
forestlands.

Contents

CHAPTER 3: SUBTRACTION / 43

CHAPTER 4: TELLING TIME / 61

CHAPTER 5: DATA AND GRAPHS / 71

CHAPTER 6: MULTIPLICATION AND DIVISION / 79

CHAPTER 7: MEASUREMENT / 93

CHAPTER 8: GEOMETRY / 103

CHAPTER 9: FRACTIONS / 131

CHAPTER 10: PROBLEM SOLVING / 145

CHAPTER 11: ANSWERS TO PRACTICE PROBLEMS / 155

CHAPTER 12: SAMPLE TEST 1 / 175

CHAPTER 13: SAMPLE TEST 2 / 203

INDEX / 231

Introduction

If you are reading this book, you are most likely planning to take the Grade 3 Mathematics Test for New York State. As a person in Grade 3, you may be taking a test of this kind for the first time. There is no need to be worried about it. This book is written to help you and your parents get ready.

The test is usually given in the spring to all Grade 3 students in New York State. The testing is done on two days. The first day you will do Book 1, which contains 40 multiple-choice questions. You will be given 60 minutes for this part of the test. The second day you will do Book 2, which contains 4 short-response questions and 2 extended-response questions. You will be asked to answer the questions and show your work in some cases, or to explain your answer. You will be given 40 minutes for this part of the test. Most students feel that this is plenty of time to do their best.

This book is designed to give you the practice and information you need to do well. Each chapter has some practice questions after some of the sections. There are also some problem-solving and extended-response questions at the end of each chapter. The solutions are listed in the answer key at the back of the book.

The last chapter of this book contains two sample tests. Take these tests after you have gone through the book so you will be well prepared. The tests are written to be just like the test you are preparing for. The format is the same with the same number of questions. You can time yourself to see how the amount of time given matches your needs.

Remember, all your teachers and your parents ask is that you do your best.

Important note: Barron's has made every effort to ensure that the content of this book is accurate as of press time, but New York State exams are constantly changing. Be sure to consult **www.p12.nysed.gov/apda/ei/eigen.html** for all the latest New York State testing information. **Regardless of the changes that may be announced after press time, this book will still provide a very strong framework for third-grade students preparing for the exam.**

Chapter 1

Place Value

Have you ever talked with your parents or grandparents about the television shows they watched when they were younger? If you have, they might have told you about a show called Name That Tune. Contestants were asked to identify popular tunes by hearing just a few notes. We are going to play Name That Number instead.

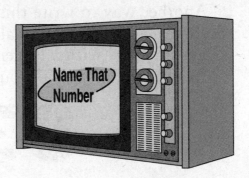

While there are different ways to write numbers, there is really only one way to read them properly. We will talk about that a little later.

How are numbers put together, and why is that important? Numbers are made up of *digits,* and each digit is in its own place. Each place has its own value, and that is what helps us to name that number. Look at the chart below.

Thousands	Hundreds	Tens	Ones
	5	2	8

This number has 5 hundreds, 2 tens, and 8 ones, so the number is called five hundred twenty-eight. Let's see how well you can name that number!

READ AND WRITE NUMBERS TO 1,000

Look at the number written below.

367

This number is written in *standard form*. In other words, the number is shown as digits. This is the way we will most often write numbers.

Another way to write this number is like this: three hundred sixty–seven. This is called *word form*.

Let's try this with a larger number.

Read this number out loud:

836

In word form this number looks like this:

eight hundred thirty-six.

Did you notice that the word "and" is never used? We will save that word to use when we are talking about money.

$5.25

We read this amount of money as five dollars and twenty-five cents. The only other time we use the word "and" is when we read a mixed number. If a recipe calls for 1½ cups of sugar, we say that we need one and one-half cups of sugar.

Let's try one more. Look at the number below and read it aloud.

972

Did you say "nine hundred seventy-two"? You were right!

Now you can see the word form!

USE EXPANDED FORM TO 999

When you study science, I'm sure you come across the word "expand." It means to get larger. When you blow up a balloon, you expand the balloon.

Did you know that a number can be expanded, too? It is called writing numbers in *expanded form,* and it looks like this:

$$972 = 900 + 70 + 2$$

If you know about place value, then you will find expanded form to be a lot of fun. In the number 972 there are 9 hundreds, or 900. There are also 7 tens, or 70, and 2 ones, or 2.

Let's try 749. There are 7 hundreds, 4 tens, and 9 ones, so the expanded form is 700 + 40 + 9.

How are you doing?

Let's try writing 509 in expanded form.

$$509 = 500 + 0 + 9$$

Did you notice what happened in the tens place? Since there are no tens, it is a good idea to put a zero in its place.

The same is true for 530 in expanded form.

$$530 = 500 + 30 + 0$$

The zero is not completely necessary, but it is better to use it.

Now let's change things around a little. Here is a number in expanded form.

See if you can change it to standard form.

$$600 + 50 + 3$$

If you said that 600 + 50 + 3 is 653 in standard form, you were right!

TEST YOUR SKILLS—READ AND WRITE NUMBERS TO 1,000 AND USE EXPANDED FORM TO 999

Circle the letter next to the best answer for each question.

1. Which of the following is 831 in word form?

   A. eighty-three

   B. eight hundred eighty-three

   C. eight hundred thirty-one

   D. eight hundred one

2. Which of the following is standard form for seven hundred fifty–nine?

   A. 75

   B. 79

   C. 795

   D. 759

3. Which of the following is standard form for 600 + 80 + 2?

   A. 600802

   B. 682

   C. 602

   D. 62

4. Which of the following is expanded form for 936?

    A. 900 + 30 + 6

    B. 960 + 3

    C. 90 + 60 + 3

    D. 9 + 3 + 6

5. Which of the following is word form for 400 + 50 + 0?

    A. four hundred five

    B. five hundred four

    C. four hundred fifty

    D. forty-five

Answers on page 155.

COMPARE AND ORDER NUMBERS TO 1,000

Have you ever caught yourself thinking you had more potato chips than your friend or fewer baseball cards than your brother or sister? You were comparing numbers of things and placing them in order. We are going to do that just with numbers. What you know about place value will help with this.

But first we need to make sure we know the signs that mathematicians use to make comparisons.

The first sign looks like this >. It means greater than. We can use it in this math sentence: 5 > 3. To read it, we say, "5 is greater than 3."

The next sign looks like this <. It means less than. Look at this sign carefully. Can you see the sideways L in it?

Think of L for less than. We can use it in this math sentence: 6 < 8. To read it, we say, "6 is less than 8." Your teachers might have taught you other ways to remember these signs.

The last sign is one you already know. It is the equal sign, and it looks like this: =. We use it to say 25 = 25. We read it by saying, "25 is equal to 25."

When there are more than two numbers to work with, you need to order them. Here's how it works.

You have been asked to order 354, 543, 453 from greatest to least.

The greatest place for each of these numbers is the hundreds place. 543 has a 5 in the hundreds place, so it is the greatest number. The next number in order is 453 because it has the next greatest hundreds place. Finally, 354 is the smallest because it is the only number left.

If you like to see the hundreds places more easily, write them in a column like this:

543
453
354

Now let's try a different list of numbers:

101, 335, 99.

You have been asked to order these numbers from greatest to least. Your eye might be drawn to 99, but is it the greatest number? No, there is a number—101—that is a little bit larger, but is that the greatest number? No! The number with the greatest hundreds place is 335, so the order from greatest to least is:

335, 101, 99.

It is easier to see their order if you write them in a column, but be sure to line the numbers up correctly.

335
101
99

PLACE NUMBERS ON A NUMBER LINE

A number line is a handy thing to use to help you order numbers. Look at the number line below.

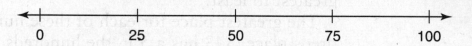

Suppose you were asked to place these numbers on the number line: 65, 20, 98, 45.

The 65 would go between the 50 and the 75. The 20 would go near the 25 but to the left. The 98 would be just to the left of the 100, and the 45 would be just to the left of the 50.

Look at the number line below.

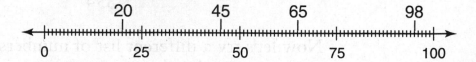

The numbers have been added for you in the right places, and if you read them from left to right, you will be reading them in order from the least to the greatest.

Let's try a new number line.

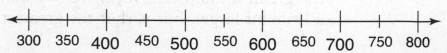

Think about where the following numbers should be placed on this new number line: 640, 460, 349, 575, 780. Now look at the number line below.

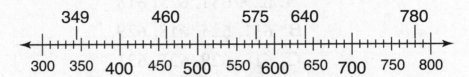

If you read the numbers from left to right you will see that from least to greatest the order is 349, 460, 575, 640, and 780.

TEST YOUR SKILLS—COMPARE AND ORDER NUMBERS TO 1,000 AND PLACE NUMBERS ON A NUMBER LINE

For questions 1–3 fill in the space as directed. For questions 4 and 5 circle the letter next to the best answer. For question 6 follow the directions in the question to place the numbers on the number line.

1. Fill in the space on the line between the numbers. Use <, >, or = to complete the number sentence.

    34 ___ 43

2. Fill in the space on the line between the numbers. Use <, >, or = to complete the number sentence.

    150 ___ 105

3. Fill in the space on the line between the numbers. Use <, >, or = to complete the number sentence.

    998 ___ 999

4. Which of the following groups is ordered from least to greatest?

A. 629, 631, 625, 618

B. 631, 625, 618, 629

C. 618, 629, 625, 631

D. 618, 625, 629, 631

5. Which of the following groups is ordered from greatest to least?

A. 899, 856, 901, 956

B. 956, 901, 899, 856

C. 956, 899, 901, 856

D. 901, 899, 856, 956

6. Place the following numbers on the number line below: 245, 423, 342

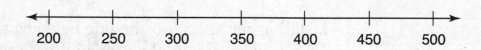

Answers on page 155.

IDENTIFY ODD AND EVEN NUMBERS

Odd and *even* numbers are easy to identify once you learn a few simple rules. Single-digit odd numbers are 1, 3, 5, 7, and 9. Odd numbers that are greater than 10 are just as easy to identify. Look at the ones place. If you see a 1, 3, 5, 7, or 9, then the number is odd. If you see a 0, 2, 4, 6, or 8 in the ones place, then the number is even. Does it seem strange that 0 is even? You know that 0 has no value, but when it is in the ones place, it can make 10. Without the zero there it would be just 1.

Let's try this out. Is 358 an odd or an even number? Look at the ones place:

358

Since 8 is even, 358 is an even number. Remember, it doesn't matter what is in the other places.

What do you think about 627? Look at the ones place:

627

Since 7 is an odd number, then 627 is an odd number.

SKIP COUNT BY 25s, 50s, AND 100s TO 1,000

When you were younger, you may have learned to count by fives. You were actually *skip counting* by five. Now that you are older, you can skip count by much larger numbers. Let's try skip counting by 100 beginning at 0.

0
100
200
300
400
500
600
700
800
900
1,000

How did you do?

Next let's try skip counting by 1,000 starting at 0.

0
1,000
2,000
3,000
4,000
5,000
6,000
7,000
8,000
9,000
10,000

Isn't this fun?

Now let's look at counting by 50. This might seem to be a little more difficult, so we will use the number line below to help us.

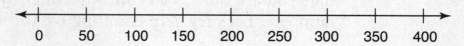

Start with your pencil at 0 and draw your skips to each mark. You should land on 0, 50, 100, 150, 200, 250, 300, 350, and end on 400. Look at the number line below to see if yours looks the same.

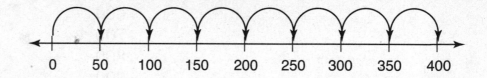

The last value we will try is skip counting by 25. This may seem strange until you think of the quarters in your piggy bank. If you can skip count by 25, you can also count your money!

Starting with no money at all, we begin at 0. Then counting quarters by 25 goes like this: 0, 25, 50, 75, 100, 125, 150, 175, 200. Do you see a pattern forming? Watch as the counting continues: 225, 250, 275, 300, 325, 350, 375, 400.

What pattern did you see? If you noticed that 0, 25, 50, and 75 repeated in each hundred, you were right.

Let's try something completely different. Look at the number line below.

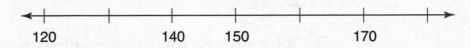

Can you figure out what goes in the blanks under this number line? Look for a hint about the skip counting. Did you notice that 140 and 150 are next to each other? What is the difference between 140 and 150? If you said 10 you were right. This number line is skip counting by 10s. The numbers that go in the blanks are 130, 160, and 180. Now look at the completed number line below.

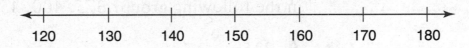

We will have more fun with numeric patterns in Chapter 10. See you there!

TEST YOUR SKILLS—IDENTIFY ODD AND EVEN NUMBERS AND SKIP COUNT BY 25s, 50s, 100s, TO 1,000

Circle the letter next to the best answer for each question.

1. Which of the following numbers is odd?

   A. 58

   B. 85

   C. 34

   D. 30

2. Which of the following numbers is even?

    A. 63

    B. 35

    C. 78

    D. 41

3. Which of the following numbers is odd?

    A. 138

    B. 250

    C. 481

    D. 572

4. When skip counting by 25, which number is missing in the following group? 375, 400, 450, 475, 500

    A. 325

    B. 425

    C. 380

    D. 420

5. When skip counting by 50, which number comes next in order?

        750, 800, 850, 900, 950, ____

    A. 775

    B. 825

    C. 925

    D. 1,000

6. Look at the number line below.

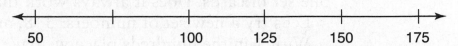

Which of the following goes in the blank under the number line?

A. 25

B. 60

C. 75

D. 90

Answers on page 156.

COMBINATIONS TO FORM 3-DIGIT NUMBERS

Three digits can be used together to form different numbers. Have you ever tried this?

Study the three cards that you see below.

How many different ways can you arrange these cards to make different numbers? If we put them in the first order, we see the number is 529. If we exchange the numbers in the tens and ones places, we will have 592. If you think there are more numbers that can be built, you are right. Move the 9 to the hundreds place, and you can make two more numbers: 952 and 925. If you place the 2 in the hundreds place, then you will have two more numbers: 259 and 295. We have made 6 new numbers with one set of cards. Does it always work that way?

Let's try a new set of numbers: 3, 8, and 1.

With 1 in the hundreds place we have 138 and 183.

With 3 in the hundreds place we have 318 and 381.

With 8 in the hundreds place we have 831 and 813.

Can you build these numbers from any other place? How about using 4, 6, and 7? Let's start with the tens place.

With 4 in the tens place we have 647 and 746.

With 6 in the tens place we have 467 and 764.

With 7 in the tens place we have 476 and 674.

PLACE VALUE PUZZLES

Before you begin each puzzle, draw ____ ____ ____ to help you.

1. Guess the number:

   - The number in my hundreds place is two more than the number in my ones place.

   - The number in my tens place is one less than the number in my hundreds place.

   - The number in my ones place is three.

2. Guess the number:

   - The number in my tens place is four more than the number in my hundreds place.

   - The number in my ones place is two more than the number in my tens place.

   - The number in my hundreds place is 1.

3. Guess the number:

- The number in my hundreds place is the same as the number in my ones place.
- The number in my tens place is three less than the number in my ones place.
- The number in my tens place is 6.

4. Guess the number:

- My digits are in counting order from the ones place to the hundreds place.
- Two of my digits are even numbers.
- My odd digit is four less than nine.

5. Guess my number:

- My hundreds digit is 3 more than my thousands digit.
- I have no ones.
- My tens digit is two more than my hundreds digit.
- My thousands digit is 4.

Answers on page 156.

PROBLEM SOLVING

1. What is $500 + 0 + 9$ in standard form?

2. Fill in the blanks in the series or pattern below:

725, _____, _____ , 800 , 825.

3. Is the number 1,458 an odd or even number? Explain how you know.

4. What is the number 698 in word form?

5. Write this number in standard form: 600 + 4,000 + 90 + 7.

6. Write all the 3-digit numbers you can make using 7, 2, and 1.

7. Order the following series of numbers from greatest to least: 918, 891, 981, 819.

8. Use <, >, or = in the space to make this number sentence true:

781 _____ 871.

9. Use the digits 5, 2, and 3 to make an even 3-digit number.

10. Explain how you know if a 2-digit number is always greater or less than a 3-digit number.

Answers on pages 156–157.

Addition

Have you ever thought about what happens when you add numbers? If there are 3 apples in your basket and you pick 4 more apples, you will have 7 apples. That is probably enough to make a great apple pie and still have some left to eat for a snack. Yum! Addition can be delicious!

Before we begin to explore addition, we should look at the parts of an addition sentence. Look at the number sentence below. Number sentences are also called *equations*.
3 + 2 = 5

3 + 2 = 5

The 3 and the 2 are called *addends*. They are the numbers you add together. Do you see the root word "add" in the word "addends"? Do you see the word "end" in the word addends? This may help you to remember it. The 5 is called the *sum*. The answer to an addition problem is always called the sum.

COMMUTATIVE PROPERTY

To help you add numbers, we will look at some *properties of addition*. The first one is the *commutative property*. This property says that the order of the addends does not change the sum. Look at this example.

$$4 + 5 = 9$$
$$5 + 4 = 9$$

You can change the 4 and 5 around, but the sum will always be 9. You may have heard this called turn around facts by teachers you had when you were younger.

Here is another way to write this.

$$4 + 5 = 5 + 4$$

This also works with bigger numbers. Take a look at this.

$$22 + 34 = 34 + 22$$

Let's check it to be sure.

$$22 + 34 = 56$$
$$34 + 22 = 56$$

It's true.

Do you think this works every time?

If you said "yes," you were right.

Let's try something. What do you think goes in the blank to make the sentence below a true addition sentence?

$$6 + 8 = \underline{} + 6$$

The commutative property says that the order of the addends does not change the sum, so the number that completes this sentence is 8.

$$6 + 8 = \underline{8} + 6$$

IDENTITY PROPERTY

Another property that helps us with addition is the *identity property*. This property says that any number added to zero remains the same number. In other words, the number keeps its identity.

Here's how it works.

$$9 + 0 = 9$$
$$0 + 4 = 4$$

Let's try some bigger numbers.

$$35 + 0 = 35$$
$$0 + 89 = 89$$

Did you already know this one?
Fill in the blank in the sentence below.

$$\underline{\hspace{2cm}} + 8 = 8$$

If you said <u>0</u>, you were right.

ASSOCIATIVE PROPERTY

The last property we are going to think about is the *associative property*. Some of you may know it as the grouping property. This property is used when you have more than two addends.

Look at the problem below.

$$3 + 2 + 6 =$$

In order to find the sum, you can group two of these addends together with parentheses.

It would look like this.

$$(3 + 2) + 6 =$$

Remember that the parentheses say, "Do me first!"

$$3 + 2 = 5$$

Now the problem looks like this.

$$5 + 6 = 11$$

What would happen if we grouped these addends another way?

$$3 + (2 + 6) =$$
$$2 + 6 = 8$$
$$So \ 3 + 8 = 11$$

The associative property says that the grouping of addends does not change the sum.

Let's try another problem.

$$4 + 2 + 7 =$$
$$(4 + 2) + 7 = \qquad 4 + 2 = 6$$
$$6 + 7 = 13$$
$$4 + (2 + 7) = \qquad 2 + 7 = 9$$
$$4 + 9 = 13$$

If we take this one step further, is the following then a true statement?

$$(4 + 2) + 7 = 4 + (2 + 7)$$

Look back at the work we just did together and you will see that this is a true math sentence.

Here is another problem.

$$5 + 2 + 4 + 8 =$$

Using the associative property, we can group the addends like this.

$$(5 + 2) + (4 + 8) =$$
$$7 + 12 = 19$$

We can also try this.

$$5 + (2 + 4) + 8 =$$
$$5 + 6 + 8 =$$
$$(5 + 6) + 8 =$$
$$11 + 8 = 19$$

Either way, we get the same sum. Pretty neat!

Here's another thing to try. Put your hands on your face so your fingers cover your cheeks. Say, "Ahhhh, associative property!" As you take your hands away from your cheeks keep them slightly bent as if they were still covering your cheeks. Do you see the parentheses they have formed?

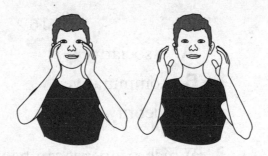

See if you can fill in the blank in the sentence below using what you know about the associative property.

$$(2 + 8) + 7 = \underline{\quad} + (8 + 7)$$

To solve this puzzle, look at the numbers on the left. You see a 2, an 8, and a 7. What do you see on the other side? You see an 8 and a 7. The 2 is missing, so it is 2 that goes in the blank.

TEST YOUR SKILLS—COMMUTATIVE PROPERTY, IDENTITY PROPERTY, AND ASSOCIATIVE PROPERTY

Circle the letter next to the best answer for each question.

1. Which property can be used to solve the number sentence below?

   $$8 + 1 = 1 + \underline{}$$

   A. associative

   B. commutative

   C. identity

2. Which property can be used to solve the problem below?

   $$16 + \underline{} = 16$$

   A. associative

   B. commutative

   C. identity

3. Which property can be used to solve the problem below?

   $$5 + (2 + 7) = (5 + 2) + \underline{}$$

   A. associative

   B. commutative

   C. identity

4. Which of the following makes the sentence below a true math sentence?

$$5 + 9 = 9 + \underline{\quad}$$

A. 5

B. 9

C. 14

D. 23

5. Which of the following makes the sentence below a true sentence?

$$0 + \underline{\quad} = 32$$

A. 0

B. 2

C. 3

D. 32

Answers on page 157.

PATTERNS IN ADDITION

When you were younger, you probably spent some time making patterns with beads or blocks. Now we are going to explore some patterns in addition. All you need to know are some basic addition facts.

3 + 5 = 8	In other words: 3 ones + 5 ones = 8 ones
30 + 50 = 80	3 tens + 5 tens = 8 tens
300 + 500 = 800	3 hundreds + 5 hundreds = 8 hundreds
3,000 + 5,000 = 8,000	3 thousands + 5 thousands = 8 thousands

Let's try another.

6 + 7 = 13	6 ones + 7 ones = 13 ones
60 + 70 = 130	6 tens + 7 tens = 13 tens
600 + 700 = 1,300	6 hundreds + 7 hundreds = 13 hundreds
6,000 + 7,000 = 13,000	6 thousands + 7 thousands = 13 thousands

What do you notice about this pattern? What is the same about each of these addition problems and what is different?

Let's look at the first example again.

$$3 + 5 = 8$$
$$30 + 50 = 80$$
$$300 + 500 = 800$$
$$3,000 + 5,000 = 8,000$$

Did you see the same basic addition fact in each addition sentence?

Now look at the zeroes. Did you see that each addend has the same number of zeroes as the sum?

Does that work every time? Try this one.

$$8 + 2 = 10$$
$$80 + 20 = 100$$
$$800 + 200 = 1000$$
$$8,000 + 2,000 = 10,000$$

Why are there more zeroes in the sum than in each addend?

There is a zero in the first sum, and that is why there is always one more zero in the sum than in each addend.

Look at this example again with the sum of ten in blue.

$$8 + 2 = 10$$
$$80 + 20 = 100$$
$$800 + 200 = 1000$$
$$8,000 + 2,000 = 10,000$$

Now you can see that the pattern works even when the sum comes with its own extra zero.

Let's try this pattern out on some new problems.

$$70 + 20 = 90$$

This works because the basic fact is $7 + 2 = 9$. Each addend and the sum have one extra zero.

$$4,000 + 8,000 = 12,000$$

This is true for the same reason.

$$500 + 500 = 1,000$$

The sum here is right because the basic fact, $5 + 5 = 10$, has a sum with a zero in the ones place.

ODD AND EVEN SUMS

In Chapter 1, we learned about odd and even numbers. Now we are going to take that one step further to learn about odd and even sums. The same rule applies to odd or even sums and to numbers. In other words, the digit in the ones place makes the sum odd or even.

Here is an example.

$$5 + 9 = 14$$

This is an even sum, because 4 is an even number.

Look at this problem again to check on the addends.

$$5 + 9 = 14$$

Both 5 and 9 are odd numbers, so in this case an odd number plus an odd number equals an even number.
Let's try another.

$$6 + 2 = 8$$

In this case both the addends and the sum are even, so an even number plus an even number equals an even number.
What other combination can you think of?
How about this?

$$7 + 4 = 11$$

Here you have an odd number, which is 7, plus an even number, which is 4. The sum is 11, which is an odd number.
Try some of these yourself to see if the rules always work.

$$8 + 9 =$$
$$6 + 4 =$$
$$9 + 3 =$$
$$11 + 5 =$$
$$12 + 2 =$$

What did you discover? Did the rules work every time? These rules are important to help you check your work.

TEST YOUR SKILLS—PATTERNS IN ADDITION AND ODD AND EVEN SUMS

Circle the letter next to the best answer.

1. Which of the following is the sum of the two addends below?

$$5,000 + 5,000 =$$

A. 10

B. 100

C. 1,000

D. 10,000

2. Which of the following is the sum of the two addends below?

$$600 + 400 =$$

A. 10

B. 100

C. 1,000

D. 10,000

3. Which of the following is the missing addend for the problem below?

$$90 + \underline{\quad} = 170$$

A. 8

B. 80

C. 800

D. 8,000

4. Which of the following sets of addends has an even sum?

A. 4 + 3

B. 5 + 7

C. 6 + 1

D. 3 + 2

5. Which of the following sets of addends has an odd sum?

A. 3 + 5

B. 8 + 2

C. 7 + 2

D. 9 + 3

Answers on page 158.

MORE THAN ONE WAY TO ADD

You may hear your parents or grandparents talk about choices in life. There are also choices in math—particularly in addition. These choices have to do with how you go about solving addition problems. Your teachers may have taught you many ways to add. Remember: When it works for you, go with it! For instance, you may have been taught to count up from the largest number in a problem. If the problem is 2 + 9, it may be easier to start with the 9 and count 2 more. But if the problem is 6 + 7, that is a different story. These numbers are called neighbors because they are next to each other when you count. Think of doubling the number on either side. You might think of 6 + 6 = 12. Since that is true, then 6 + 7 is one more because 7 is one more than 6, so 6 + 7 = 13. If you thought of the double 7 + 7 = 14, then 6 + 7 is one less because 6 is one less than 7. Look at the problems below to help you.

$$6 \quad 6 \quad 7$$
$$\underline{+6} \quad \underline{+7} \quad \underline{+7}$$
$$12 \quad 13 \quad 14$$

Do you see how the sums are neighbors too?

Students often find adding 9 difficult until they learn this little trick. When adding 9 to a number, think of the 9 as 10, add it to the other addend, and then take away one. Look at the examples below.

$$10 \quad 9$$
$$\underline{+5} \quad \underline{+5}$$
$$15 \quad 14$$

Here's another way to add 9. Take one away from the other addend and add the resulting number to 10.
Look at the examples below.

$$9 \quad 10$$
$$\underline{+8} \quad \underline{+7}$$
$$17 \quad 17$$

ADDING 3-DIGIT NUMBERS

Now that you are getting comfortable with adding single or small double-digit numbers, let's look at larger numbers with 3 digits. We will be thinking about regrouping or carrying when necessary. Let's try some!

$$345 + 96$$

Add the numbers in the ones places.
Regroup if necessary.

$$\overset{1}{}$$
$$345$$
$$\underline{+96}$$
$$1$$

In this case, 5 + 6 is 11, so the 1 in the ones place is written directly under the 6, and the 1 in the tens place is carried or regrouped above the 4 in the tens place.

Add the numbers in the tens places.

Regroup if necessary.

$$
\begin{array}{r}
\overset{1\,1}{3\,4\,5} \\
+\,9\,6 \\
\hline
4\,1
\end{array}
$$

In this case, 4 + 9 plus the 1 that was carried or regrouped has a sum of 14, so the 4 is placed directly under the 9 in the tens place, and the 1 is carried or regrouped above the 3 in the hundreds place.

Add the numbers in the hundreds places.

$$
\begin{array}{r}
\overset{1\,1}{3\,4\,5} \\
+\,9\,6 \\
\hline
4\,4\,1
\end{array}
$$

Now try one with two 3-digit addends.

$$478 + 931$$

$$
\begin{array}{r}
\overset{\,1\,\,}{4\,7\,8} \\
+\,9\,3\,1 \\
\hline
1{,}4\,0\,9
\end{array}
$$

Remember to regroup only when necessary. In this problem the sum of the numbers in the ones places did not need regrouping but the sum of the numbers in the tens and hundreds places did.

TEST YOUR SKILLS—MORE THAN ONE WAY TO ADD AND ADDING 3-DIGIT NUMBERS

Remember to rewrite each problem vertically before solving!

1. 683 + 925

2. 719 + 456

3. 837 + 939

4. 269 + 874

5. 378 + 609

Answers on pages 158–159.

ESTIMATING SUMS

This section may be the most important part of this chapter. As you become an older person and no longer go to school, most of the math you do will involve estimating. *Rounding* is a big part of estimating.

There is a rule for rounding, and there is also a picture for you to carry in your mind to help you round. Think of a series of 10 hills and valleys. Look at the drawing below.

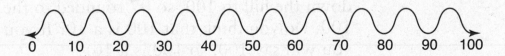

This is a picture of a number line that has been folded as if to make a fan. Did you notice that the numbers are counting by 10 beginning at 0? Now we are going to add something to this number line in order to work with it.

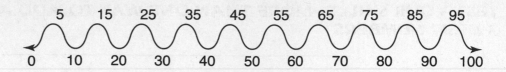

Look at the numbers at the top of each hill. Do you see that they are just to the right of the top and could slide down that hill to the right every time? We are going to use this drawing to round numbers to the nearest 10.

Remember, the numbers at the tops of the hills will always slide down to the valley to the right.

Put the number 18 on the number line above. If 18 is a marble it will roll down the hill and end up in the valley of 20. So we can say that 18 rounded to the nearest 10 is 20.

Now try the 43 on the number line. This number will roll down the hill into the valley of 40. 43 rounded to the nearest 10 is 40.

How about 64? Even though this number is high up on the hill, it will roll into the valley of 60, so 64 rounded to the nearest 10 is 60.

See if you remember this. What is 45 rounded to the nearest 10? You were right if you said 50. Remember that 45 sits to the right of the top of the hill and falls to the right.

Here is a funny one. What is 4 rounded to the nearest 10? If you place it on the hills and valleys, you will see that it will roll into the valley of 0, so 4 rounded to the nearest 10 is 0!

What do you think about 97? This number will roll down the hill to 100, so 97 rounded to the nearest 10 is 100. Do you think that 100 is a 10? If you count by 10s you will say 100, so 100 is a 10.

Now let's look at rounding to 100. The drawing below should look familiar to you. The lines are the same but the numbers are different.

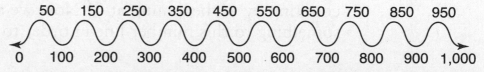

Place the number 367 on the number line. If you get it in the right spot, you will see that it will roll down into the valley of 400. It is not really possible to be exactly right in where you place the number, but so long as it is on the right side of the correct hill, you are good to go.

How about the number 599? This one doesn't have very far to roll, but it will round to 600.

Here's the rule that goes with this mind picture.

Find the place you are rounding to.
Look one place to the right.
If it is 4 or less, the rounding number stays the same.
If it is 5 or more, the rounding number goes up one.

The number of zeroes in the rounded number matches the number you are rounding to. For instance, all numbers rounded to the nearest 10 have at least one zero in them.

TEST YOUR SKILLS—ESTIMATING SUMS

Write your answer on the line provided with each question. Use your hills and valleys pictures to help you. Put your answer on the line next to the question.

---

1. What is 58 rounded to the nearest 10? _____

2. What is 61 rounded to the nearest 10? _____

3. What is 334 rounded to the nearest 100? _____

4. What is 928 rounded to the nearest 100? _____

5. What is 251 rounded to the nearest 100? _____

Answers on page 159.

Now we are ready to actually estimate some sums. An *estimate* is a number that is close to the exact answer. Most often, an estimate contains a lot of zeroes because estimates are sums of rounded addends.

Here is an example.

658	→	700
+843	→	+800
		1500

658 rounded to the nearest 100 is 700

843 rounded to the nearest 100 is 800

Remember: When estimating a sum, round each addend to the greatest place and add.

Try this one.

462	→	500
+84	→	+80
		580

ADDING COINS AND BILLS USING $0.00 TO INDICATE SUMS

Adding money is a very important skill. Hopefully you will always be adding to your money!

Suppose you have 3 quarters, 2 dimes, and 4 pennies. How much money is that?

Each quarter is 25¢, each dime is 10¢, and each penny is 1¢.

Three quarters are 75¢, two dimes are 20¢, and 4 pennies are 4¢.

$$75¢ + 20¢ + 4¢ = 99¢$$

If you find a penny, then you will have 100¢. Another name for 100¢ is $1.00.

If you find a nickel, you will now have $1 and 5¢. That is written like this:

$1.05.

This is the decimal system of writing money. You will learn more about decimals later.

Let's add some more money. The only thing to remember is to line up the decimal points.

$15.61 + $4.59 will look like this.

$$
\begin{array}{r}
1\ 1 \\
\$15.61 \\
+4.59 \\
\hline
\$20.10
\end{array}
$$

TEST YOUR SKILLS—ADDING COINS AND BILLS USING $0.00 TO INDICATE SUMS

1. What is the sum of $32.09 + $6.81? _____

2. What is the sum of $25 + $1.95? _____

3. What is the sum of $16.45 + $14.55? _____

4. What is the sum of $18.95 + $11.05? _____

5. What is the sum of $7.23 + $12.49? _____

Answers on page 159.

PROBLEM SOLVING

Circle the letter next to the best answer.

1. Which of the following is an example of the associative property of addition?

   A. 2 + 3 = 3 + 2

   B. 8 + 0 = 8

   C. 5 + 4 = 9

      4 + 5 = 9

   D. 4 + (3 + 2) = (4 + 3) + 2

2. Which of the following is an example of an odd number plus an even number being equal to an odd sum?

   A. 7 + 3 = 10

   B. 5 + 2 = 7

   C. 3 + 5 = 8

   D. 2 + 4 = 6

3. Millie has 4 coins in her pocket with a value of $0.36. Two of the coins are the same. What coins does Millie have in her pocket?

   A. 2 quarters, a dime, and 1 penny

   B. 2 dimes, 2 nickels, and 1 penny

   C. 1 quarter, 2 nickels, and 1 penny

   D. 1 quarter, 2 dimes, and 1 penny

4. Which of the following is an even number?

    A. 36

    B. 85

    C. 33

    D. 27

5. Which of the following is the best estimate of the sum of 451 + 239?

    A. 800

    B. 700

    C. 690

    D. 600

6. Which property is shown in 9 + 0 = 9?

    A. associative

    B. commutative

    C. identity

    D. estimation

7. Using what you know about patterns in addition, which of the following is the sum of 60 + 40?

    A. 10

    B. 100

    C. 1,000

    D. 10,000

8. Which of the following is shown by 2 + 8 = 10?

    A. odd plus even equals odd

    B. even plus odd equals odd

    C. even times even equals even

    D. even plus even equals even

9. Which of the following is an example of the commutative property of addition?

    A. $5 + (9 + 2) = (5 + 9) + 2$

    B. $6 + 7 = 7 + 6$

    C. $3 + 0 = 3$

    D. $4 + 6 = 10$

10. Which of the following is 372 rounded to the nearest 10?

    A. 300

    B. 360

    C. 370

    D. 400

Answers on pages 159–160.

EXTENDED-RESPONSE QUESTIONS

Emma went shopping one Saturday. She spent $3.95 on earrings, $4.50 on socks, and $8.95 on a hat.

A. Explain on the lines below how you would **estimate** how much Emma spent.

_____

_____

_____

_____

B. What is the **exact** amount she spent? Show your work in the space below.

C. She paid with a $20.00 bill. How much change did she receive? Show your work in the space below.

Answers on page 160.

<table>
<tr><td>

Chapter
3

</td><td>

Subtraction

</td></tr>
</table>

As a younger person you may have come to dread the words "take away" or subtraction. Actually, there are four ways to look at it. One is a comparison, as in "How much older are you than your brother or sister?" or "What is the difference in your ages?" The word *difference* is what we call the answer to a subtraction problem. Also, if you eat two carrot sticks and have 3 carrot sticks left, you have subtracted from the number you had to start with. Subtraction can be delicious, too! Sometimes you need to find out the number of things in a group. There are 10 sandwiches on a plate and 3 of them are ham sandwiches. How many are not ham?

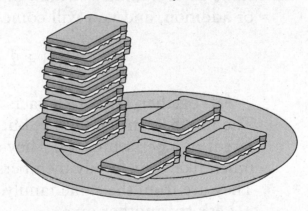

Finally, you may need to find out how many more are needed. This most often happens with money. You need $15 to buy a shirt, and you have earned $8. How much more money is needed to buy the shirt?

FACT FAMILIES

Did you know that addition and subtraction are related to each other?

Look at the subtraction fact below.

$$5 - 3 = 2$$

You can exchange the 3 and the 2 so that you have a new fact.

$$5 - 2 = 3$$

Using the same three numbers, we can make an addition sentence.

$$2 + 3 = 5$$

Can you think of another addition sentence using the same three numbers? Think of the commutative property of addition, and you will come up with this.

$$3 + 2 = 5$$

All together, these four facts are called a *fact family*. The numbers are the same, but they are placed in four different number sentences. They are *related* to each other much like your family members are related to each other. They are from the same family, but each is different.

Let's try another one.

This time we will use 8, 7, and 1.

$$7 + 1 = 8$$
$$1 + 7 = 8$$
$$8 - 1 = 7$$
$$8 - 7 = 1$$

Did you notice something special about the subtraction sentences? They always begin with the greatest number!

Let's try one more. We will use 9, 12, and 3.

$$12 - 9 = 3$$
$$3 + 9 = 12$$
$$12 - 3 = 9$$
$$9 + 3 = 12$$

You do not have to write them in any particular order. Just make sure that each one is different.

Here is a special fact family. We will use 5, 10, and 5. Did you notice that two members of this family are twins? Do you think that makes a difference? Let's try it!

$$5 + 5 = 10$$
$$10 - 5 = 5$$

That's it! Twins make all the difference! Twin fact families have only two members.

PATTERNS IN SUBTRACTION

You may have heard a teacher or one of your parents tell you how important it is to know your basic facts. They were right! Watch what you can do with some basic subtraction facts.

If you know that

$$7 - 5 = 2,$$

then you know that

$$70 - 50 = 20$$

and

$$700 - 500 = 200.$$

What do you notice about these number sentences?

They all contain the basic fact that we started with:

$$7 - 5 = 2.$$

What pattern did you see in the other number sentences? Did you notice the zeroes? Each time we see the same number of zeroes in the difference as in each of the numbers being subtracted.

Let's try a crazy big number to see if it works.

$$70,000 - 50,000 = 20,000$$

There were four zeroes in each of the numbers being subtracted, and there were four zeroes in the difference. Let's try something else. Here is a special basic fact.

$$10 - 6 = 4$$

It is special because there is a zero in the largest number. Remember to add the same number of zeroes to each number, and the pattern will work.

$$100 - 60 = 40$$
$$1000 - 600 = 400$$

Can you see that the basic fact is still there?

$$\underline{1}000 - \underline{6}00 = \underline{4}00$$

TEST YOUR SKILLS—FACT FAMILIES AND PATTERNS IN SUBTRACTION

Circle the letter next to the best answer.

1. Which of the following is a related fact of 7 – 3 = 4?

    A. 3 + 4 = 7

    B. 7 – 5 = 2

    C. 7 + 3 = 10

    D. 4 + 7 = 11

2. 8 + 6 = 14 is related to which of the following facts?

    A. 6 + 2 = 8

    B. 14 – 5 = 9

    C. 8 – 2 = 6

    D. 6 + 8 = 14

3. Which of the following belongs to the 7, 15, 8 family?

    A. 8 – 7 = 1

    B. 8 + 7 = 15

    C. 15 – 9 = 6

    D. 7 + 5 = 12

4. Which of the following subtraction sentences is the result of the basic fact 12 – 4 = 8?

    A. 40 – 10 = 30

    B. 80 = 40 = 40

    C. 120 – 90 = 30

    D. 120 – 40 = 80

5. Which of the following basic facts could result in this difference?

$$200 - 160 = 40$$

A. $20 - 18 = 2$
B. $20 - 16 = 4$
C. $16 - 8 = 8$
D. $16 - 4 = 12$

Answers on pages 160–161.

REGROUPING IN SUBTRACTION

There comes a moment in every student's life when the teacher says, "You can't do that!" Unfortunately, this statement sometimes comes up in subtraction. Let's set the record straight! You actually <u>can</u> subtract a larger number from a smaller number, but that is a lesson for another day. For now, we will regroup the numbers we are working with to make it possible to subtract smaller numbers from larger ones.

Look at the problem below.

$$\begin{array}{r} 72 \\ -6 \\ \hline \end{array}$$

Beginning with the ones place, the problem reads 2 − 6. We will need to get some help for the 2 in order to subtract 6 from it. Get out some base ten blocks and build the number 72 using 7 tens rods and 2 units cubes. If you don't have base ten blocks, use the pictures shown on the next page. Exchange one of the rods for 10 units cubes. You still have the number 72, but now you have 12 units cubes to work with and 6 tens rods left.

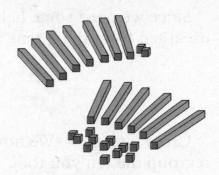

Take 6 units cubes off your desk, or cross them off in this book, leaving 6 units cubes. The 6 tens rods remain, so the number you have as a difference is 66.

Let's try another problem.

$$83$$
$$-37$$

Once again, looking at the ones column, we have the problem 3 − 7 so we need to go to the tens for help. Build the number 83 with base ten blocks. Exchange one of the tens rods for 10 units cubes. Now you have 7 tens rods and 13 units cubes. This still has a value of 83.

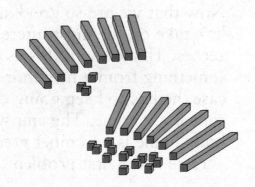

Since 13 − 7 is 6, put the 6 under the 7. Now subtract 3 from 7 to get 4.

With the 4 under the 3, the difference of 83 − 37 is 46.

Now let's try one without the base ten blocks.

$$95$$
$$-78$$

Since we need some help to subtract 8 from 5, we will use a ten from the 9 tens in 95.

$$\begin{array}{r} 8\ 15 \\ \cancel{9\ 5} \\ -78 \\ \hline \end{array}$$

Cross out the 9. We now have 8 tens. Exchange or regroup the ten you took from the 9 for ten ones. Add 10 ones to 5 to make 15 ones.

Now the problem is easily solved as 15 − 8 is 7 and 8 − 7 is 1. The answer is 17.

$$\begin{array}{r} 8\ 15 \\ \cancel{9\ 5} \\ -78 \\ \hline 17 \end{array}$$

Remember!!! Only regroup when you have to! If the bottom number is less than the top one, you are good to go without regrouping!!!

REGROUPING ACROSS ZEROES

Now that we are so good at regrouping in subtraction, let's take on the most interesting task of regrouping across zeroes. Think about this as someone who needs to borrow something from a neighbor but no one is home. In that case, he has to keep going down the street until he finds somebody home. The empty homes are zeroes, and anybody home is a number greater than zero.

Here is the first problem.

$$\begin{array}{r} 600 \\ -357 \\ \hline \end{array}$$

In the ones column the problem reads 0 − 7. We need some help from the tens, but nobody is home. We go next door to the hundreds and find a 6. We take one hundred from the 6 to make it 5. Now we move to the tens and add 10 to the zero. We must borrow a ten, so 10 becomes

9. The ones column now has 10 to work with. Build this with base ten blocks to check it. Take out 6 hundreds flats. Trade one in for 10 tens rods to make 5 hundreds flats and 10 tens rods. Now trade in one tens rod for 10 units cubes. Check to make sure this still has a value of 600.

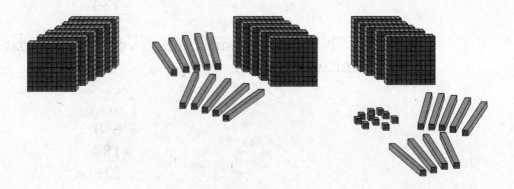

Look at the way the problem is written below.

$$\begin{array}{r} 9 \\ 5\ \cancel{10}\ 10 \\ \cancel{6}\ \cancel{0}\ \cancel{0} \\ -357 \end{array}$$

Now the problem is all set up for you to solve. In the ones column, 10 − 7 = 3. In the tens column 9 − 5 = 4, and in the hundreds column 5 − 3 = 2.

$$\begin{array}{r} 9 \\ 5\ \cancel{10}\ 10 \\ \cancel{6}\ \cancel{0}\ \cancel{0} \\ -357 \\ \hline 243 \end{array}$$

Go back to the base ten blocks and do the problem with them. Did you get the same answer?

Let's try one on our own without the blocks.

$$\begin{array}{r} 400 \\ -184 \\ \hline \end{array}$$

Move to the hundreds place and take 1 hundred to make 3. There is now 10 above the tens place. Regroup that to make 9 tens and 10 ones. Here is how it should look.

$$
\begin{array}{r}
9\\
3\ \cancel{10}\ 10\\
\cancel{4}\ \cancel{0}\ \cancel{0}\\
-184\\
\end{array}
$$

Now do the subtraction you have prepared to get your answer.

$$
\begin{array}{r}
9\\
3\ \cancel{10}\ 10\\
\cancel{4}\ \cancel{0}\ \cancel{0}\\
-184\\
\hline
216\\
\end{array}
$$

How did you do? Let's try another problem.

$$
\begin{array}{r}
800\\
-358\\
\end{array}
$$

Regroup as you need to in order to solve the problem.

$$
\begin{array}{r}
9\\
7\ \cancel{10}\ 10\\
\cancel{8}\ \cancel{0}\ \cancel{0}\\
-358\\
\hline
442\\
\end{array}
$$

TEST YOUR SKILLS—REGROUPING IN SUBTRACTION AND REGROUPING ACROSS ZEROES

1. $$
\begin{array}{r}
81\\
-53\\
\end{array}
$$

2. $$
\begin{array}{r}
63\\
-35\\
\end{array}
$$

3.   58
       −29

4.   300
       −136

5.   600
       −374

Answers on page 161.

ESTIMATING DIFFERENCES

If you like rounding, you will like this lesson. As you get older you will do more and more estimating. As a matter of fact, as an adult you will use estimating and mental math most of the time you do any math. Look at the problem below.

Emma has 78 coins in her collection, and her friend, Millie, has 21 coins. About how many more coins does Emma have than Millie?

78
−21

This is not a difficult subtraction problem to find the actual answer for, but there is a magic word in the problem that changes the way you must do it. That magic word is "about!" When you see "about" in any problem, you know you need not give the actual answer. Instead, round each number to the greatest place first and then subtract.

$$78 \rightarrow 80$$
$$-21 \rightarrow -20$$
$$60$$

Emma has about 60 more coins than Millie.

What happens if the numbers are bigger? In a way it is even easier as the numbers get bigger. Read the problem below.

There were 495 people who attended the school play on Friday night. On Saturday afternoon 246 people attended. <u>About</u> how many more people attended the play on Friday night?

Set the problem up and then round to the greatest place.

$$495 \rightarrow 500$$
$$\underline{-246 \rightarrow -200}$$
$$300$$

<u>About</u> 300 more people attended on Friday night.

One of the most common uses for estimating is when we use money.

Shane has saved $36.78, and Ben has saved $48.53. <u>About</u> how much more has Ben saved than Shane?

$$\$48.53 \rightarrow \$50$$
$$\underline{-36.78 \rightarrow -40}$$
$$\$10$$

Ben has <u>about</u> $10 more than Shane.

ODD OR EVEN DIFFERENCES

Once you have solved a subtraction problem, it is important to be able to see if your answer is reasonable. One way to do that is to know about even and odd differences. Let's try a few.

To find the difference of an even number minus and odd number we will try 8 – 5. The difference is 3, which is an odd number.

To find the difference of an even number minus an even number we will try 6 – 4. The difference is 2, which is an even number.

To find the difference of an odd number minus an even number we will try 11 – 6. The difference is 5, which is an odd number.

To find the difference of an odd minus an odd number we will try 7 – 3. The difference is 4, which is an even number.

Let's see how that works.

$$\begin{array}{r} 96 \\ -45 \\ \hline 51 \end{array}$$

Since 96 is an even number and 45 is and odd number, then it is reasonable that the answer of 51 is an odd number.

SUBTRACTING 3-DIGIT NUMBERS

Now that you have practiced subtraction with smaller numbers, let's try some 3-digit numbers.

$$\begin{array}{r} 973 \\ -642 \\ \hline \end{array}$$

Beginning with the ones column, 3 – 2 = 1. In the tens column 7 – 4 = 3 and 9 – 6 = 3 for a difference of 331. Did you notice that you didn't have to regroup this time? Let's try something a little more challenging.

$$\begin{array}{r} 832 \\ -547 \\ \hline \end{array}$$

Beginning with the ones column, we see the need to regroup from the tens and add ten to the ones to make 12.

$$\begin{array}{r} {\scriptstyle 2\ 12} \\ 8\cancel{3}\cancel{2} \\ -547 \\ \hline 5 \end{array}$$

Now we have trouble in the tens column! We need to regroup a hundred to help out the tens.

$$\begin{array}{r} \overset{12}{7\,\cancel{2}\,12} \\ \cancel{8}\,\cancel{3}\,\cancel{2} \\ -547 \\ \hline 285 \end{array}$$

Whew! That was a real challenge!

What happens when a zero creeps in? We worked with subtracting across zeroes earlier in this chapter, but look at this problem.

$$\begin{array}{r} 603 \\ -358 \\ \hline \end{array}$$

There is only one zero. Beginning with the ones column, we see the need to regroup, but nobody is home in the tens column. We will have to regroup in the hundreds. Adding 10 tens to the zero leaves 10 in the tens place. Now that 10 can be regrouped, leaving 9 in the tens column and 13 in the ones column.

$$\begin{array}{r} 9 \\ 5\,\cancel{10}\,13 \\ \cancel{6}\,\cancel{0}\,\cancel{3} \\ -358 \\ \hline 245 \end{array}$$

TEST YOUR SKILLS—ESTIMATE DIFFERENCES, ODD OR EVEN DIFFERENCES, AND SUBTRACTING 3-DIGIT NUMBERS

Circle the letter next to the best answer for each question.

1. Which of the following will result in an even difference?

    A. 10 – 3

    B. 5 – 2

    C. 11 – 5

    D. 8 – 1

2. Which of the following is an estimate of the difference of 66 – 47?

A. 10

B. 19

C. 20

D. 25

3. Which of the following is the difference of 756 – 437?

A. 319

B. 321

C. 329

D. 400

4. Which of the following will result in an odd difference?

A. 15 – 11

B. 9 – 7

C. 6 – 2

D. 8 – 3

5. Which of the following is an estimate of the difference of 89 – 32?

A. 40

B. 50

C. 57

D. 60

Answers on page 162.

PROBLEM SOLVING

1. There were 863 marbles in a bag. Mattye spilled 345 marbles out of the bag. About how many were left in the bag?

2. A card store sold 598 birthday cards and 856 other cards in two months. How many more other cards did the store sell in that time?

3. Molly practices the piano 145 minutes each week. So far this week, she has practiced 105 minutes. How much longer does she have to practice this week?

4. Zoe and her younger brother have a difference of 5 years in their ages. If Zoe is 9 years old now, how old is her brother?

5. There are sixty-one students who play sports in South High School. Twenty-three of them play more than one sport. How many of the students play only one sport?

6. Philip has a goal to collect 100 baseball cards. So far, he has collected 37 baseball cards. How many more does he need to collect to reach his goal?

7. A video store has 900 videos to rent. Last weekend all but 176 videos were rented. How many videos were rented?

8. Susan has 387 books on her bookshelves. She has 165 books about horses. About how many books are not about horses?

9. George has a collection of 56 matchbox cars. Only 27 of them are in a box. How many of George's cars are not in a box?

10. There are 46 players in the band and 29 of them are girls. There are 6 drummers. How many boys are in the band?

Answers on page 162.

EXTENDED-RESPONSE QUESTIONS

1. Ben practiced the guitar double the minutes that Shane practiced the piano. If both students practiced a total of 60 minutes, how long did each student practice his instrument? Show your work in the space below.

2. Mr. Frank drives a bus in the city. This morning 15 people got on his empty bus at the first stop. Then 6 people got off at the next stop. At the third stop, 8 people got on his bus. How many people were riding Mr. Frank's bus then? Show your work in the space below.

3. Wade went shopping last Saturday. He bought socks for $1.45 and a shirt for $6.55. How much change did Wade receive for $10? Show your work in the space below.

Answers on page 163.

Chapter 4

Telling Time

Have you ever thought about the different kinds of clocks there are? There are tall grandfather clocks and small alarm clocks and many medium–sized clocks. These are called *analog* clocks.

There are also *digital* clocks. They are electric and mostly small clocks that can sit on a table.

Analog clocks have a round face with the numbers 1 to 12 on them arranged so that the 12 is at the top, the 6 at the bottom and the other numbers evenly divided between them in order. This kind of clock can show only 12 hours at a time. What do you think is wrong with that? There are more than 12 hours in a full day. As a matter of fact there are exactly twice that number, or 24 hours, in a full day. We use A.M. and P.M. to help identify which 12 hours we are talking about. For example, we eat breakfast at 7 o'clock in the morning, or 7 A.M. We eat our dinner at 6 o'clock in the evening, or 6 P.M.

TELLING TIME TO THE HOUR

When you were younger you most likely learned how to tell time by the hour. We will review this quickly. Study the clock face shown below.

    This clock shows 5 o'clock because the hour hand, the shorter one, is on the five and the minute hand, the longer one, is on the 12.
    Whenever the hour hand is on the hour and the minute hand is on the 12, the time is on the hour.

TELLING TIME TO THE HALF-HOUR

Telling time on the half-hour is a little tricky because of the position of the hour hand. The minute hand is on the 6 and the hour hand is halfway between two numbers. Study the clock below.

    This clock shows 3:30 because the minute hand is on the 6 and the hour hand is halfway between the 3 and the 4.

TEST YOUR SKILLS—TELLING TIME TO THE HOUR AND TELLING TIME TO THE HALF-HOUR

Circle the letter next to the best answer.

1. Which of the following clocks shows 8:30?

A. 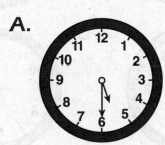 **B.**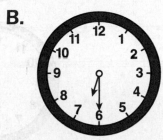

C. 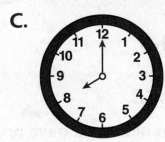 **D.**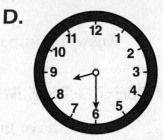

2. Which of the following is the time shown on the clock face below?

A. 6:30

B. 7:00

C. 7:30

D. 8:00

3. Which of the following clocks shows 12:30?

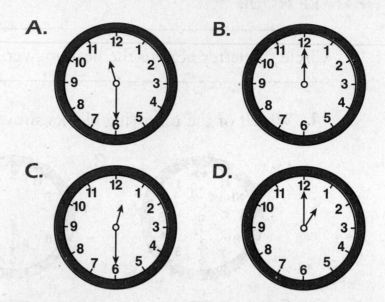

A. B.

C. D.

Answers on page 163.

TELLING TIME TO THE MINUTE

As you have grown older you have probably noticed that telling time does not always work out on the hour and half-hour. Let's look at how to use all those numbers in between. As the minute hand moves around for every hour, it is important to know that every number it passes stands for 5 minutes. So when the minute hand moves from the 1 to the 2, for example, five minutes have passed. In order to help you remember this, the clock below has been drawn with the minute numbers just outside the face of the clock.

Study the clock below.

The hour hand is between the 10 and the 11, and the minute hand is on the 4. The time shown is 10:20. We say ten twenty. We can also say 20 minutes after 10.

Remember: The hour is always the number that the hour hand was on last, or in most cases the lower number. Let's try another. Study the clock below.

The time shown here is 3:40. We say three forty or 40 minutes after 3. Since this time is past the half-hour, we can also say 20 minutes before 4. Look back at the clock face and notice that the hour hand is slightly past the halfway point between the hours of 3 and 4.

Here's one that will be even more fun. Study this clock carefully.

The time shown is 11:23. We say eleven twenty-three or 23 minutes after 11.

USING $\frac{1}{2}$ AND $\frac{1}{4}$ TO TELL TIME

Sometimes fractions are used to help us tell time. Look at the clock face drawn below.

Shade the section to the right of the line; you have shaded half of the clock. When the minute hand passes from the 12 to the 6 it has passed through half the hour, so 4:30 is also half past 4.

Now look at the clock face below.

This clock has been divided into four equal pieces. Shade the piece in the top right to show how much time the minute hand has passed through from the 12 to the 3.

Now the clock has been divided into quarters or fourths, so we can say that the time is 8:15 or a quarter past 8.

If the time is 4:45, the minute hand has passed through three-fourths or three-quarters of the hour. Instead of saying three-quarters past the hour, we now say a quarter to 5 or a quarter to the next hour.

TEST YOUR SKILLS—TELLING TIME TO THE MINUTE

1. Which of the following clocks shows 5:50?

A.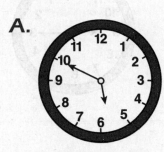

B.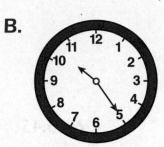

C.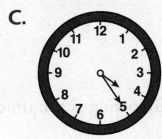

D.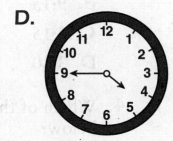

2. Which of the following is the time shown on the clock below?

A. 9:15

B. 3:09

C. 3:15

D. 3:45

3. Which of the following is the time shown on the clock below?

A. 3:45

B. 9:15

C. 3:15

D. 3:30

4. Which of the following is the time shown on the clock below?

A. 7:50

B. 10:07

C. 7:45

D. 10:35

Answers on page 163.

PROBLEM SOLVING

1. Maria waited 30 minutes for her bus. How long is that in quarter-hours?

2. Erica is eating lunch with her friends. Which of the following is most likely the time?

   A. 3:15 P.M.

   B. 12:05 A.M.

   C. 5:30 A.M.

   D. 12:15 P.M.

3. Alika is meeting her friends at the movies at 1:15 P.M. Which of the following clocks is showing that time?

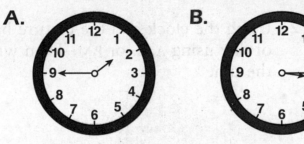

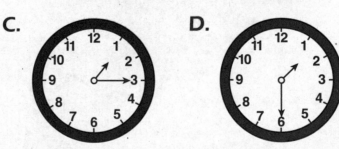

4. Austin is eating breakfast at 7:45 A.M. What is another way of saying seven forty-five?

5. Describe in words the position of the hour hand when the time is 8:30.

Answers on page 164.

EXTENDED-RESPONSE QUESTIONS

1. Using the clock and the picture below, write the time of day using A.M. or P.M. Then write one way to read the time.

2. Using the clock and the picture below, write the time of day using A.M. or P.M. Then write one way to read the time.

3. Jeff's baseball game is set to begin at 3:35 P.M. Is this closer to 3:00 P.M. or to 4:00 P.M.? Tell how you know.

Answers on page 164.

Chapter 5

Data and Graphs

Ginny is having a birthday party and has invited her friends. Along with the cake, she is planning to have ice cream and she wants all her friends to have the flavors they like. She asked each friend to tell her his or her favorite flavor. In order to be sure she buys enough of each flavor, Ginny can put this *data* she *collected* in a *tally chart* and *frequency table*.

Here is the list of her friends and the flavors they like.

Kathy—chocolate
Bethany—butter pecan
Liz—vanilla
Anita—chocolate
Heidi—vanilla
Deanna—chocolate
Libby—butter pecan
Kara—chocolate
Kelly—chocolate
Mary—vanilla

Flavor	Tally	Total
Chocolate	~~IIII~~	5
Butter pecan	II	2
Vanilla	III	3

The chart above shows the tally marks for each flavor and the number of times or frequency of each flavor. Now

Ginny can be sure she has enough of each flavor of ice cream for her friends.

PICTOGRAPHS

After you have collected some data and made a frequency chart, you can graph the data. A graph is a way of displaying data. If Ginny wanted to graph the data she collected, she could make a *pictograph*. Study the graph below.

Favorite Flavors of Ice Cream for Ginny	
Chocolate	☺☺☺☺☺
Butter pecan	☺☺
Vanilla	☺☺☺

Key: ☺ = 1 vote

This is a pictograph of Ginny's data. The *key* is under the graph and tells you how many votes each picture is worth. Don't forget to check the key before you read a graph!

Here is some data collected by Kurt when he asked his classmates about their favorite television shows. It is shown in a frequency table.

Type of Show	Number of Votes
Sports	10
Game shows	6
Cops	8
Cartoons	3

Key [ ] = 2 votes

Key [= 1 vote

The pictograph for this data could look like this.

Favorite Television Shows	
Sports	▢ ▢ ▢ ▢ ▢
Game shows	▢ ▢ ▢
Cops	▢ ▢ ▢ ▢
Cartoons	▢ ▢

Key: ▢ = 2 votes

Key: ▢ = 1 vote

BAR GRAPHS

Another graph form is the *bar graph*. This graph displays data in the form of bars that can be vertical (up and down), or horizontal (side to side).

Alex collected data on favorite summer activities. His frequency table is below.

Activity	Number of Votes
Camping	10
Swimming	12
Hiking	6
Reading	8
Gardening	2

The bar graph for Alex's data could look like this.

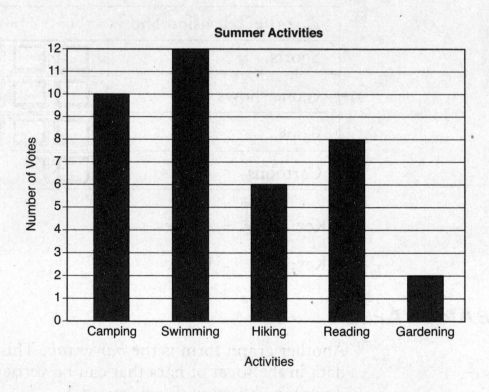

SCALE

Actually, this graph could be placed on a grid that is much smaller. In this case we change the *scale*, so that each line is counted by 2s.

It would look like this.

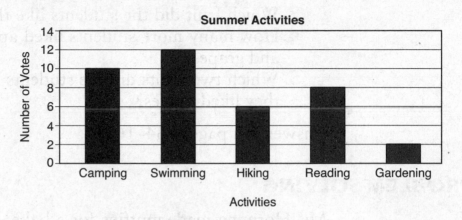

READING AND INTERPRETING GRAPHS

Once you have made a graph, you can read and interpret it by comparing the bars or pictures. Graphs make data easy to compare. The tallest or longest bar, or the most pictures, represents the largest amount of data. The shortest or smallest bar, or the fewest pictures, represents the least amount of data. Study the bar graph below.

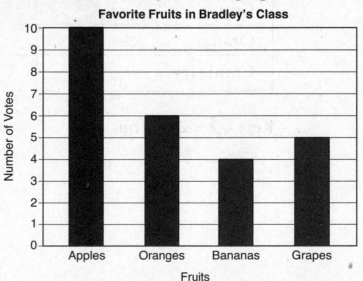

TEST YOUR SKILLS—HERE ARE SOME QUESTIONS YOU CAN ANSWER USING THIS GRAPH.

1. Which fruit did Bradley's class like the most?
2. How many more students liked oranges than bananas?
3. Which fruit did the students like the least?
4. How many more students liked apples than bananas and grapes?
5. Which two fruits did the students like as much as they liked apples?

Answers on pages 164–165.

PROBLEM SOLVING

Ms. Hornung made muffins for a bake sale. She made 16 blueberry, 10 corn, 6 orange, and 8 cranberry muffins. Complete the pictograph below using this data.

MS. HORNUNG'S MUFFINS

Type of Muffin	Number of Muffins
Blueberry	
Corn	
Orange	
Cranberry	

Key: ◯ = 2 muffins

Joey is collecting marbles in a jar. The table below shows how many of each color Joey has.

MARBLES IN JOEY'S JAR

Color	Number Collected
Red	8
Yellow	6
Blue	12
Purple	10

Complete the bar graph below using the data in the table of marbles in Joey's jar.

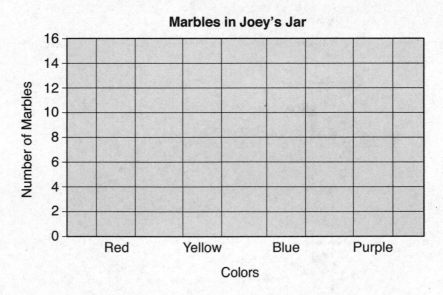

Answers on page 165.

Multiplication and Division

When you were younger, you learned about addition and subtraction. You learned how to use them and when to use them. Now that you are older, you will learn more about multiplication and division. Once you try these new operations, you will like them. They are faster ways to add and subtract.

MULTIPLICATION IS ALSO REPEATED ADDITION

Here is how multiplication works. Study the multiplication problem below.

$$2 \times 4 = 8$$

The 2 and the 4 in this number sentence are called *factors*. They are the numbers that do the work of multiplying. The 8 is called the *product*. The answer to any multiplication problem is the product.

This problem can be read in two ways.

2 groups of 4 are 8.
2 times 4 is 8.

Here's what that means.

$$4 + 4 = 8$$

So far this isn't any faster than addition. Now let's look at a different problem.

$$3 \times 4 = 12$$

This problem is also 4 + 4 + 4 =12.
The first factor tells you how many times to add the second factor.
Study this problem.

$$5 \times 4 = 20$$

This time the addition looks like this.

$$4 + 4 + 4 + 4 + 4 = 20$$

Try this problem.

$$6 \times 2 = 12$$

The addition problem is 2 + 2 + 2 + 2 + 2 + 2 = 12

MULTIPLICATION AS AN ARRAY

An *array* is an arrangement of dots or small pictures that are used to show a multiplication sentence.
3×4 looks like this as an array.

There are 3 rows of 4 that make a total of 12 apples.

3 × 4 is also 3 groups of 4.
Let's try another one.

$$5 \times 3 = 15$$

It is important to remember that we are talking about rows that go across and columns that go up and down.

TWO DIFFERENT PROBLEMS WITH THE SAME PRODUCT

This is one of the magical things that occur in multiplication. Here's how it works.
2 × 3 = 6 looks like this as an array.

$3 \times 2 = 6$ looks like this as an array.

These two arrays look different and are really two very different problems, but they have the same product. They each belong to a different multiplication table. Actually, this is one thing that makes learning your multiplication facts a little easier.

TEST YOUR SKILLS—MULTIPLICATION IS REPEATED ADDITION, MULTIPLICATION AS AN ARRAY, AND TWO DIFFERENT PROBLEMS WITH THE SAME PRODUCT

Follow the directions in each question. Use the space provided for your answers.

1. Circle the product in the problem below.

$$6 \times 3 = 18$$

_____

_____

2. Draw an array for $7 \times 2 = 14$.

_____

_____

_____

_____

3. Write the addition problem for $5 \times 3 = 15$.

_____

_____

4. Study the number sentence below.

$$2 \times 5 = 10$$

Write another multiplication problem with the same product but using the same numbers.

_____

_____

5. Write an addition problem for $8 \times 2 = 16$.

_____

_____

Answers on page 166.

COMMUTATIVE PROPERTY

You may remember the commutative property from your study of addition. This property works the same way for multiplication. The commutative property says that the order of the factors does not change the product. You may have already noticed this when you read about the arrays $3 \times 2 = 6$ and $2 \times 3 = 6$.

Now we can put this together, and we have

$$3 \times 2 = 2 \times 3.$$

The commutative property allows us to find missing factors. Study the problem below.

$$5 \times 4 = \underline{\hspace{1cm}} \times 5$$

The number that makes this a true number sentence is 4 because $5 \times 4 = 20$ and $4 \times 5 = 20$, so 4 is the missing number.

Try this one.

$$6 \times 7 = 7 \times \underline{\hspace{1cm}}$$

The answer is 6 because 6 is the missing number. You don't have to know that $6 \times 7 = 42$.

IDENTITY PROPERTY

The identity property of multiplication says that whenever you multiply by 1 you get a product of the other factor. Here's how it works.

$$9 \times 1 = 9$$
$$1 \times 5 = 5$$

This happens with bigger numbers, too.

$$23 \times 1 = 23$$
$$1 \times 56 = 56$$

Let's try some.

$$1 \times 18 =$$

If you said 18, you were right!

$$48 \times 1 =$$

The answer is 48.
How great is this?

ZERO PROPERTY

The zero property says that any number multiplied by zero has a product of zero.

Here's how this one works.

$$0 \times 4 = 0$$
$$8 \times 0 = 0$$

This also works with larger numbers.

$$38 \times 0 = 0$$
$$0 \times 96 = 0$$

Let's try some of these.

$$73 \times 0 =$$

If you said the product is 0, you were right.

$$0 \times 29 =$$
$$87 \times 0 =$$
$$0 \times 63 =$$

The product for all of these number sentences is 0!

TEST YOUR SKILLS—COMMUTATIVE PROPERTY, IDENTITY PROPERTY, AND ZERO PROPERTY

Circle the letter next to the best answer.

1. Study the number sentence below.

$$45 \times 1 = 45$$

This is an example of which of the following properties?

A. commutative

B. identity

C. zero

2. Study the number sentence below.

$$8 \times 9 = 9 \times 8$$

This is an example of which of the following properties?

A. commutative

B. identity

C. zero

3. Study the number sentence below.

$$6 \times 0 = 0$$

This is an example of which of the following properties?

A. commutative

B. identity

C. zero

4. Study the number sentence below.

$$7 \times 5 = 5 \times \underline{\quad}$$

According to the commutative property, which of the following numbers makes this a true number sentence?

A. 7

B. 5

C. 35

5. Study the number sentence below.

$$89 \times 1 = \underline{\quad}$$

According to the identity property, which of the following numbers make this a true number sentence?

A. 1

B. 89

C. 90

Answers on page 166.

PRACTICING BASIC MULTIPLICATION FACTS

The moment of truth arrives. Multiplication has been introduced to you, and now you must learn all these facts on your own! You look at the pages of facts you have written, and maybe you have made flash cards that are stacked up on your desk. Now what?

The best thing you can do is get someone your age or older to help you. A study buddy your own age might be someone you ride the bus with. Every morning and every afternoon use that time to study facts. You can use flash cards or quiz each other out loud.

An older person in your home can help you as well. Set up a time each day to work on facts. Do this for no more than 15 minutes every day.

Perhaps you go shopping with one of your parents every day. Use the time in the car to study facts. Ask your mom or dad to quiz you.

If you are using flash cards at home, remove the cards from the pile as you learn them. You will see the stack of facts you know grow taller and the stack of facts you don't know get smaller.

There are also some good websites for practicing multiplication facts. Ask your teacher to suggest some.

You can use any one or more than one of these methods to help you learn your facts. The most important thing is to do something every day to practice. You will be truly amazed to see how quickly you get some good results. All of a sudden you will realize that it is much easier to do math problems at school because you have learned your facts.

The best way to learn these is with someone. Going to your room alone is not a good idea!

Remember: The only person that is really responsible for learning your facts is you! The more you practice, the better off you will be.

DIVISION IS ALSO REPEATED SUBTRACTION

Division is a word that sometimes scares students. Think of it this way. Division is related to subtraction the same way multiplication is related to addition. In other words, division is repeated subtraction.

Study the division problem below.

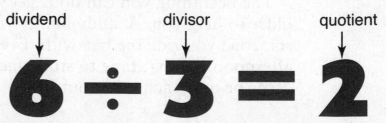

dividend   divisor   quotient

$$6 \div 3 = 2$$

$$6 \div 3 = 2$$

$$3\overline{)6}^{\,2}$$

The *dividend* is the number being divided. The *divisor* is the number doing the work or doing the dividing. The *quotient* is the answer.

Here's how it works.

$$10 \div 2 = 5$$

We start with 10 and subtract 2 from it until we get to zero.

$$10 - 2 = 8$$
$$8 - 2 = 6$$
$$6 - 2 = 4$$
$$4 - 2 = 2$$
$$2 - 2 = 0$$

There are 5 blue 2s because $10 \div 2$ is 5. In other words, there are 5 groups of 2 in 10.

Let's try another one.

$$18 \div 6 = 3$$
$$18 - 6 = 12$$
$$12 - 6 = 6$$
$$6 - 6 = 0$$

There are 3 red 6s because 18 ÷ 6 is 3. There are 3 groups of 6 in 18.

DIVISION-MULTIPLICATION FACT FAMILIES

At the beginning of Chapter 3 we learned about addition-subtraction fact families. These two operations work together because they are *opposite* or *inverse* operations. The same can be said about multiplication and division. In this chapter we learned that multiplication is repeated addition and that division is repeated subtraction. Let's take a closer look at this idea.

We know that $2 \times 3 = 6$ and $3 \times 2 = 6$ because of the commutative property. Using the same three numbers, we can make division sentences like this.

$$6 \div 3 = 2 \text{ and } 6 \div 2 = 3$$

If we put this all together we have a new kind of fact family.

$$2 \times 3 = 6$$
$$3 \times 2 = 6$$
$$6 \div 3 = 2$$
$$6 \div 2 = 3$$

Let's try another one.
We will use the numbers 9, 2, and 18.

$$9 \times 2 = 18$$
$$2 \times 9 = 18$$
$$18 \div 2 = 9$$
$$18 \div 9 = 2$$

Did you notice that the greatest number always begins division problems in the same way that the greatest number always began subtraction problems?

Here is the key to learning division facts. Once you know your multiplication facts, all you need to do is turn them around and you have a division fact! Your knowledge of fact families is a powerful tool!

PROBLEM SOLVING

1. Three workers load fifteen boxes on a truck. If each worker loads the same number of boxes, how many does each worker load?

2. Josephine bought a pencil at the school store. She paid for it with 3 nickels. How much did the pencil cost?

3. Kristin is buying candy to share with friends. She bought 6 pieces that cost 4¢ each. How much did Kristin pay for the candy?

4. Mackenzie puts 5 T-shirts in each of 4 drawers. How many T-shirts does Mackenzie have in all 4 drawers?

5. Tennis balls come three in a can. If Roger buys 24 tennis balls, how many cans of balls does he have?

Answers on page 167.

EXTENDED-RESPONSE QUESTIONS

1. Spencer put three pictures on each of 5 pages in his album. Gawain put five pictures on each of 3 pages in his album. Which person had more pictures? Explain your answer in words or number sentences.

2. At Darby's camp there were 3 canoes to take 2 people each and 4 rowboats to take 3 people each. How

many people could ride in the canoes and rowboats altogether? Show your work in the space provided.

3. Mrs. Cosgrove's class has 3 violins that have 4 strings each. If her class gets 2 more violins, how many strings will they have then? Show your work in the space provided.

4. Cali has a new photo album that holds 30 photos. If 6 friends give her 4 photos each, how many more will she need to fill her new album? Show your work in the space provided.

5. Mr. Kneeland asked his 28 students to arrange themselves in 4 rows with the same number in each row. Draw a picture of this problem and write a number sentence to solve it.

Answers on page 167.

Measurement

It's Monday morning and you open your pencil box. You realize that your pencils are getting shorter and your erasers are almost all gone. With your eyes, you were measuring the things in your pencil box. Measurement is something we can do using rulers and other tools. After using these tools for a while you will find yourself estimating some measurements.

LINEAR MEASUREMENT—CUSTOMARY

Let's begin with *line*ar measurement. Do you see the word "line" in the word "linear"? That will help you remember that linear measurement is distance measurement or how long something is. The customary system is the system we inherited from the settlers in this country who came from England. It is a system put together by kings and not mathematicians, so it has some strange parts.

Look at your ruler!

Remember: Some rulers begin with 0, and some don't have numbers until 1.

When you are measuring how long something is, start at the first line on the ruler.

WHOLE INCHES AND HALF-INCHES

The *inch* is the smallest measurement in this system. Since it is not very small, we have to break it up into fractions of inches. For now, we will stick to *whole inches* and *half-inches*. Look at the picture of the pencil on the next page.

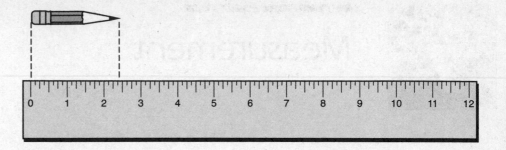

This pencil is just a little shorter than $2\frac{1}{2}$ inches long. If you are asked to measure it to the nearest half-inch, the pencil is about $2\frac{1}{2}$ inches long. If you are asked to measure it to the nearest inch, it is about 2 inches long because it measures less than $2\frac{1}{2}$ inches. If it was a little longer than $2\frac{1}{2}$ inches, then it would measure about 3 inches long. Let's try measuring some other things. Use your ruler to measure each of these items and write your answer on the line next to each item. Measure to the nearest half-inch first and then measure to the nearest whole inch.

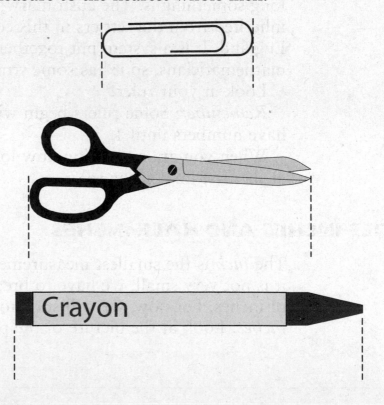

The paper clip is just a little longer than $1\frac{1}{2}$ inches, so it is about $1\frac{1}{2}$ inches long to the nearest half-inch. It is also about 2 inches long to the nearest whole inch.

The scissors are a little less than 3 inches long. This one is a little tricky. The scissors are about 3 inches long. That is actually your answer both times. Remember it takes two half-inches to make a whole inch!

The crayon is a little longer than $3\frac{1}{2}$ inches, so it is about $3\frac{1}{2}$ inches to the nearest half inch. It is also about 4 inches long to the nearest whole inch.

FEET

The next largest measurement in the customary system is the *foot*. The story is that a king's foot was measured and its length became the foot we use to this day. A foot is 12 inches long.

Medium-size distances are measured in feet. The length of your bedroom is a certain number of feet. The length of a hallway in a home is measured in feet. The driveway up to your house may be measured in feet. Many rulers are exactly one foot or 12 inches long. If you put your thumbs together and spread your hands you will have about 1 foot.

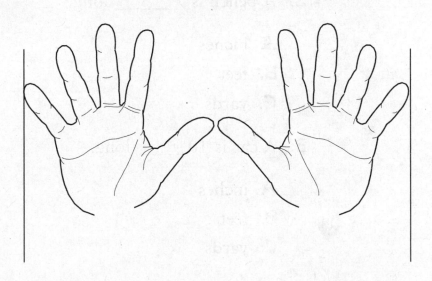

YARDS

The next largest measurement in the customary system is the *yard*. There are 3 feet in a yard, and there are 36 inches in one yard. If you have ever played football or watched a football game, you know that a football field is measured in yards. We use yards to measure slightly longer things. The sidewalk in front of your school can be measured in yards. The main hallway in your school can be measured in yards.

TEST YOUR SKILLS—LINEAR MEASUREMENT (CUSTOMARY), WHOLE INCHES AND HALF-INCHES, FEET, AND YARDS

Circle the letter next to the best answer.

1. A desk is 25 _____ wide.

    A. inches

    B. feet

    C. yards

2. A pencil is 7 _____ long.

    A. inches

    B. feet

    C. yards

3. A car is 10 _____ long.

    A. inches

    B. feet

    C. yards

4. A baseball field is 100 _____ long.

    A. inches

    B. feet

    C. yards

5. A book is 15 _____ wide.

    A. inches

    B. feet

    C. yards

Answers on page 168.

WEIGHT

The customary system has two common measures for weight. The lighter one is the *ounce*. The heavier one is the *pound*. There are 16 ounces in each pound. If you go shopping for food with your family, you may see someone buy meat or fruits and vegetables by the pound. A Thanksgiving turkey could weigh as much as 20 pounds. That makes for a lot of turkey leftovers! A watermelon could weigh as much as 15 pounds. That's a lot of refreshment on a hot summer day!

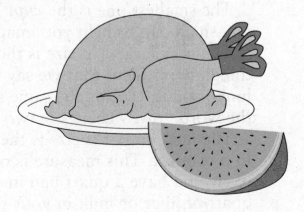

Some things that could be measured in ounces are a baseball, a football, a small bag of potato chips, a box of ten pencils, and a CD.

Some things that could be measured in pounds are your desktop computer, your wide-screen television, a textbook, your backpack, and even you!

CAPACITY

The measure of the amount of liquid in a container is called its *capacity*. There are four different measures of capacity that we will explore.

The smallest one is the *cup*. You can think of almost any small cup to help you imagine how much it can hold.

The next largest in size is the *pint*. The only really tricky thing about this is that we say the word "pint" with a long i sound. This is the same sound we use when we say the word "my." There are two cups in one pint.

The next largest is size is the *quart*. There are two pints in one quart. This measure is often used in cooking. Most likely you have a quart pan in your kitchen and maybe a quart of juice or milk in your refrigerator. The word "quart" is the root word found in the word "quarter."

One-quarter is another way of saying one-fourth. One quart is actually one-fourth of the next larger measure of capacity.

The *gallon* is the largest measure of capacity we will study. It takes four quarts to make a gallon. Gallons are used for large containers like swimming pools. Milk often comes in gallon containers. There may be some sitting in your refrigerator right now!

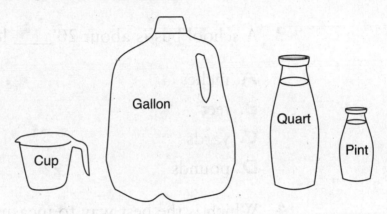

TEST YOUR SKILLS—WEIGHT AND CAPACITY

Circle the letter next to the best answer.

1. Which of the following is likely to weigh 5 pounds?

    A. laptop computer

    B. CD

    C. cell phone

    D. wristwatch

2. Which is the best way to measure the length of your computer keyboard?

    A. inches

    B. feet

    C. yards

    D. pounds

3. A school bus is about 20 _____ long.

    A. inches

    B. feet

    C. yards

    D. pounds

4. Which is the best way to measure the water in your bathtub?

    A. cups

    B. pints

    C. quarts

    D. gallons

5. Which is the best way to measure the amount of juice in a juice box?

    A. cups

    B. pints

    C. quarts

    D. gallons

Answers on page 168.

ESTIMATING MEASUREMENTS

As an older person, you will find yourself estimating measurements. Here's how it works. A teacher asks you to take a note to another classroom. She tells you that the room you want is about 15 feet down the hall. You don't have a ruler to use to measure the distance, so you must estimate or take a good guess. Estimation is used for capacity as well. The next time you watch a cooking show on television, listen to hear the chef tell you to use about 2 quarts of water to cook the pasta. If you listen carefully, you will hear the word "about" used more often than you might expect when it comes to measurement.

If a chicken weighs just a little less than 3 pounds, we say that it weighs about 3 pounds. Remember: Just like rounding, if the item being estimated is more than half of the next measure, we go up. If it is less than half of the next measure it stays the same.

PROBLEM SOLVING

1. A box of nails weighs 3 pounds. How much does 2 boxes weigh?

2. A desk is 1 yard wide, and a table is 3 feet wide. Which one is wider?

3. Sam's backyard is 9 yards wide. How many feet wide is Sam's backyard?

4. Measure the line below to the nearest inch.

_____

5. Maria bought 2 gallons of milk. Katherine bought 2 quarts of milk. Who bought more milk? Explain your answer.

Answers on page 169.

EXTENDED-RESPONSE QUESTIONS

1. Milly is measuring the length of the wall of her bedroom to see if her new bookcase will fit. Which unit of measure should she use? Explain your answer on the lines below.

2. Jaden has a piece of string that is one foot long. How many pieces can Jaden cut from this string that are exactly 4 inches long? Hint: Draw a picture to help you.

3. Damien has $1\frac{1}{2}$ pounds of candy to share with his friends. How many ounces of candy does he have? (1 pound = 16 ounces.) Show your work in the space below.

Answers on page 169.

Chapter 8

Geometry

Have you ever felt like you were surrounded on all sides? Well, when you really think about it, geometry is everywhere you look! It's not as bad as it sounds. Geometry adds a lot of beauty to our lives and sometimes even makes life easier! The most obvious is called *solid* geometry or *three-dimensional* geometry. Any object that you can pick up or that has length, width, and height is three-dimensional.

The three-dimensional or solid figures we are going to study are: *cone, cylinder, cube, prism*, and *sphere*.

The *face* of a solid figure is any flat surface of the figure. An *edge* of a solid figure is the line segment formed where two faces meet. A *vertex* of a solid figure is either the point where three or more edges meet or the tip of a cone. If you have more than one vertex, they are called *vertices*. Look at the pictures of examples below:

Examples:

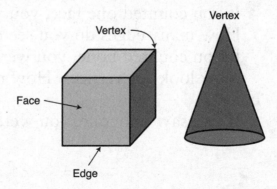

How does all this work with each solid figure? Let's try it out!

THREE-DIMENSIONAL OR SOLID FIGURES

Here is a cone.

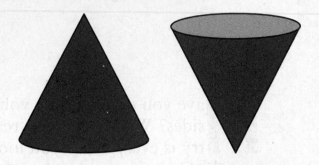

Look at the examples of cones you might find in real life.

How many faces can you see on one of these cones?
If you counted one face, you were right.
How many edges do you see on any one of these cones?
If you counted none, you were right.
Now look for vertices. How many do you see on a cone?
If you saw only one, you were right!

Here is a cylinder.

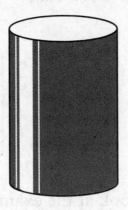

Here are some examples of cylinders you might see every day.

How many faces does a cylinder have?
If you counted two, you were right.
What about edges? Can you find any?
How about vertices?
If you didn't count any vertices or edges, you were right. How do you think this happens?
Actually, the faces never meet, and that is why there are no edges and no vertices.

Here is a cube.

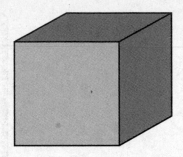

Look at the examples of cubes you might find in your life.

Can you see the faces? How many do you count? Are some of the faces hidden from you in this picture? How many are there altogether?

If you counted six faces, you were right.

How many edges are there? Are some of them hidden from you in this picture, too? How many edges are there altogether?

It you counted 12 edges, you were right.

Have you noticed some vertices? How many are there, including the ones you can't see, in this picture?

If you said eight vertices, you were right.

Here is a rectangular prism. Look at the picture.

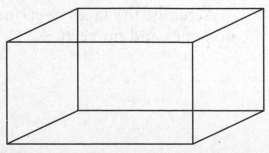

How do you think this prism got its name?

Can you find some rectangles? That's where the name came from!

Here are some rectangular prisms you might see every day.

How many faces can you count, including the ones that are hidden, in this picture?

If you found six faces, you were right.

How many edges does each rectangular prism have?

If you said 12 edges, you were right.

How many vertices can you count on each rectangular prism?

If you said eight vertices, you were right.

Here is a sphere. Look at the picture.

The sphere is a lot of fun to study because you have probably used spheres in your life since you were very young. What do you think is the most popular form of a sphere in your life? Look below to find some hints.

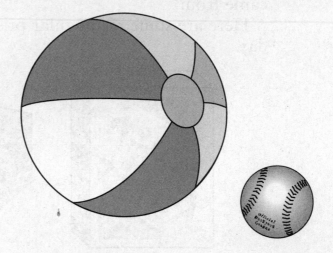

How many faces do you see on any of these spheres?
If you said none, you were right.
How many edges can you count?
Right! There are none again!
Can you find any vertices?
Of course not! If there are no faces or edges, there can be no vertices either.
Can you think of the most important sphere in your life?
Look at the drawing below.

This is the sphere on which we all live. This is the earth.
Look at the table below showing the faces, edges, and vertices for all the solid figures we have studied.

Solid Figure	Faces	Edges	Vertices
Cone	1	0	1
Cylinder	2	0	0
Cube	6	12	8
Rectangular prism	6	12	8
Sphere	0	0	0

Did you notice that the cube and the rectangular prism have the same number of faces, edges, and vertices? Do you know why?

A cube is really a special kind of rectangular prism. We will talk more about this later when we study polygons.

TEST YOUR SKILLS—THREE-DIMENSIONAL OR SOLID FIGURES

Circle the letter next to the best answer for each question.

1. Which of the following is not a word used to describe a solid figure?

   A. vertex

   B. face

   C. angle

   D. edge

2. A tree trunk is closest to which solid figure?

   A. sphere

   B. cone

   C. cube

   D. cylinder

3. Which of the following has only one face and one vertex?

   A. sphere

   B. cone

   C. cylinder

   D. cube

4. Which two solid figures have the same number of faces, edges, and vertices?

   A. sphere and cone

   B. cone and cylinder

   C. cube and cone

   D. cube and rectangular prism

Answers on pages 169–170.

Now we are going to study a part of geometry called *plane geometry* because it is two-dimensional. The figures have only length and width. You may be able to pick up the piece of paper that a figure is drawn on, but the figure itself has only two dimensions.

POLYGONS

A *polygon* is a closed figure whose sides are all line segments.
   Study the chart below that shows the polygons we will be studying and the number of line segments it takes to make them.

Name of the Polygon	Number of Line Segments
Triangle	3
Square	4
Rectangle	4
Rhombus	4
Trapezoid	4
Pentagon	5

A *triangle* is a polygon with three sides.

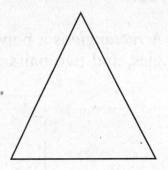

Triangles often take their special names from an angle used to form them. For example, the triangle drawn for you below is called a right triangle because it has one right angle shown with a little box in it.

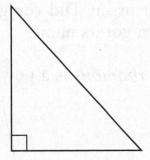

A *square* is a polygon with four right angles and four equal sides.

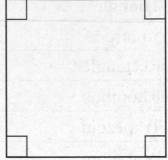

A square is the polygon that is the shape of each face of a cube.

A *rectangle* is a polygon with four sides, four right angles, and two pairs of equal sides.

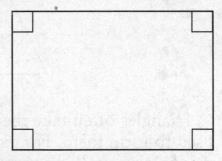

A rectangle is the shape of most of the faces of a rectangular prism. Did you guess that that's how a rectangular prism got its name?

A *rhombus* is a polygon with four equal sides.

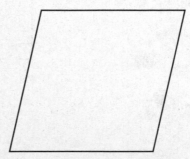

A *trapezoid* is a polygon with four sides. A trapezoid also has only one pair of parallel sides.

Look at the figure above. Did you notice that two of the sides are darker than the others? Those are the parallel sides.

A *pentagon* is a polygon with five sides.

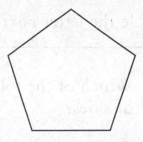

There is a very famous building in Washington, D.C. If you were a bird flying over this building, it would look a lot like this.

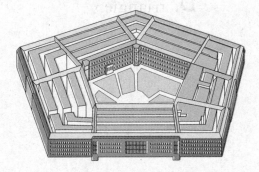

The last figure we will study is the circle. It is not a polygon because line segments do not form it. Actually, a path of points around a center point forms a circle. Study the circle drawn below.

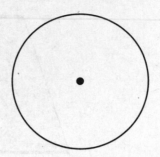

The circle is the shape of both faces of a cylinder and the only face of a cone.

TEST YOUR SKILLS—POLYGONS

Circle the letter next to the best answer for each question.

1. Which of the following polygons has the least number of sides?

   A. pentagon

   B. square

   C. rhombus

   D. triangle

2. Which polygon has been drawn below?

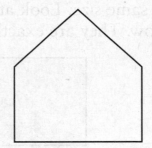

A. square
B. rectangle
C. pentagon
D. rhombus

3. Which of these is **not** a polygon?

A. circle
B. rhombus
C. triangle
D. pentagon

4. Which of these is a trapezoid?

A.　　　　　　　　　**B.**

C.　　　　　　　　　**D.**

Answers on page 170.

CONGRUENT AND SIMILAR FIGURES

Two figures are *congruent* if they are the same shape and the same size. Look at the drawings of the two rectangles below. They are exactly the same, so they are congruent.

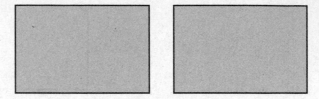

Look at the drawings of the rectangles below. These are also congruent even though one is turned a little.

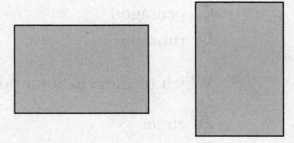

Two figures are *similar* if they are the same shape but not necessarily the same size. Look at the trapezoids drawn below.

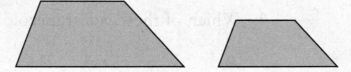

Did you notice that there is magic in geometry? Some of that magic is in congruent and similar figures because all congruent figures are also similar. The same thing is not true about all similar figures.

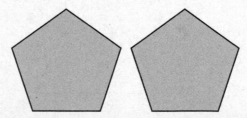

The pentagons above are congruent and similar because they are the same size and the same shape.

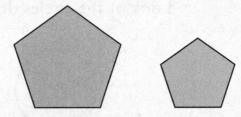

This second set of pentagons are similar but not congruent because they are the same shape but not the same size. Another important thing to remember is that just because two figures have the same name, they are not always similar. Look at the two triangles below.

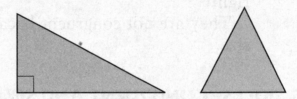

Even though these are both triangles, they are not similar.

Study the squares below.

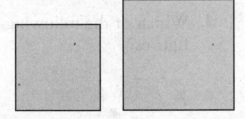

If you think these squares are similar, you are right. Now look at the squares below.

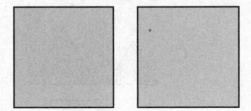

These squares are congruent because they are exactly the same. They are also similar.

Have you thought about circles?

Look at the circles drawn below.

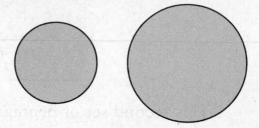

Do you think these two drawings are congruent, similar, or both?

If you think they are similar but not congruent, you are right!

They are not congruent because they are not the same size.

TEST YOUR SKILLS—CONGRUENT AND SIMILAR FIGURES

Circle the letter next to the best answer for each question.

1. Which of the following is an example of two congruent figures?

A.

B.

C.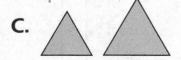

D.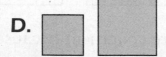

2. Which of the following figures are similar and congruent?

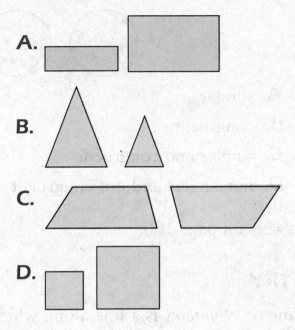

3. Which of the following best describes the figures below?

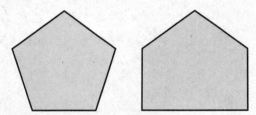

A. similar

B. congruent

C. similar and congruent

D. not similar and not congruent

4. Which of the following best describes the figures below?

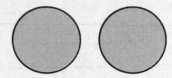

A. similar

B. congruent

C. similar and congruent

D. not similar and not congruent

Answers on page 170.

LINE OF SYMMETRY

A *line of symmetry* is a line along which a figure can be folded so that the two halves match exactly. Look at the heart shape drawn below.

Have you ever made a Valentine's Day card? If you have, you may remember folding a paper in half, tracing a half heart on the fold and cutting it out. Like magic, a full heart unfolds with a line of symmetry down the center.

Let's look at some of the polygons we have studied to see about lines of symmetry for them.

Some triangles have a line of symmetry. Look at the one drawn below.

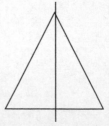

How many lines of symmetry do you think a square has? Look at the drawing below.

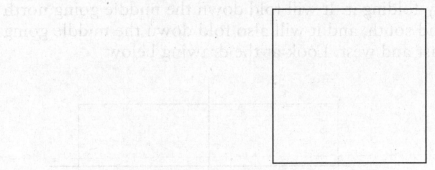

Draw some lines of symmetry of your own on this square, or cut one out from another piece of paper and try to fold it.

You should discover that there are four lines of symmetry on a square. The drawing below shows them all.

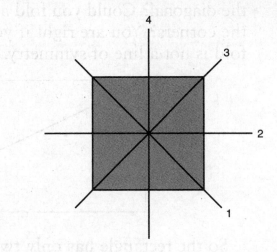

Now that you have done the square, how many lines of symmetry do you think the rectangle has? Look at the rectangle drawn for you below.

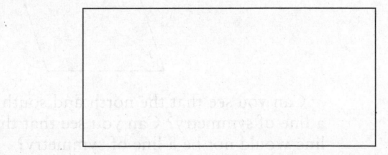

Did you think that there should be four lines of symmetry for the rectangle, too? Take a regular piece of paper and try folding it. It will fold down the middle going north and south, and it will also fold down the middle going east and west. Look at the drawing below.

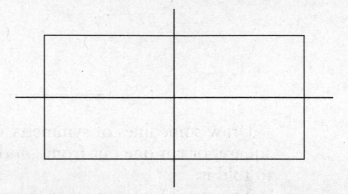

What happened when you tried to fold the paper along the diagonal? Could you fold a diagonal and still match all the corners? You are right if you found that the diagonal fold is not a line of symmetry.

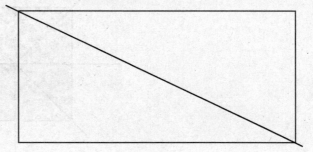

So the rectangle has only two lines of symmetry. How about the rhombus?

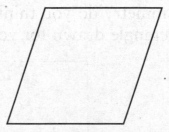

Can you see that the north and south line would not be a line of symmetry? Can you see that the east and west line would not be a line of symmetry?

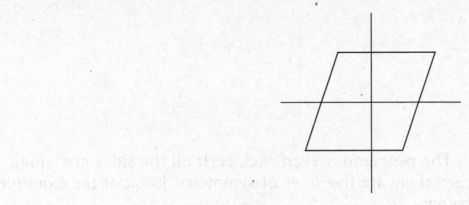

Now try drawing the diagonal lines. They will both work as lines of symmetry. The rhombus has two lines of symmetry.

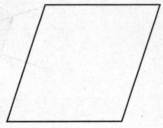

The trapezoid is a tricky polygon. If the two sides that are not parallel are equal, then that trapezoid will have one line of symmetry down the middle. Look at the drawing below.

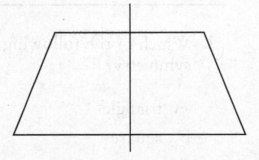

Other trapezoids will not have a line of symmetry. Here's one for you to look at on the next page.

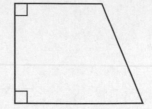

The pentagon is even trickier. If all the sides are equal, then there are five lines of symmetry. Look at the example below.

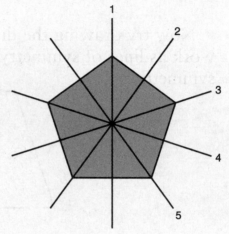

TEST YOUR SKILLS—LINE OF SYMMETRY

Circle the letter next to the best answer for each question.

1. Which of the following polygons has the most lines of symmetry?

 A. triangle

 B. square

 C. regular pentagon

 D. rhombus

2. How many lines of symmetry does a rectangle have?

    A. 1
    B. 2
    C. 3
    D. 4

3. How many lines of symmetry does this drawing of a tree have?

    A. 4
    B. 3
    C. 2
    D. 1

4. How many lines of symmetry does this figure have?

    A. 0
    B. 1
    C. 2
    D. 3

Answers on page 170.

PROBLEM SOLVING

Use the lines below each question to write your answers.

1. How are a cube and a rectangular prism alike?

2. How are a cube and a rectangular prism different?

3. Why does a sphere have no edges or vertices?

4. Sam painted a color cube for a game he made. If he used a different color for each side of the cube, how many colors did Sam need?

5. Jamisen took two cubes of wood and glued them together face to face to make one solid shape. What solid shape did she make?

6. Cooper has two blocks that are different but both have the same circular face. Which two solid shapes does Cooper have?

7. Why does a square have four lines of symmetry and a rectangle have only two lines of symmetry?

8. How many lines of symmetry does a circle have?

9. How are a square and a rhombus the same?

10. How are a square and a rhombus different?

11. Why are a parallelogram and a triangle not similar?

_____

12. What makes all circles similar?

_____

Answers on page 171.

EXTENDED-RESPONSE QUESTIONS

Look at the figures drawn for you below.

_____   _____   _____

1. Write the name of each figure on the line below it. On the lines provided below write two ways to identify a rectangle.

_____

_____

2. Identify the objects below by writing the solid geometric name on the lines below each object.

_____   _____   _____

On the lines below write one thing these shapes have in common.

_____

_____

3. Draw lines of symmetry on each of the figures below.

_____   _____   _____

Write the number of lines of symmetry on the lines below each figure.

4. On the grid below, draw three squares. Two of the squares must be congruent, and the third must be similar to the other two squares.

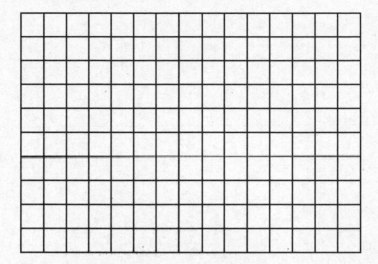

On the lines below, write two reasons why all squares are similar.

_____

_____

_____

Answers on pages 171–172.

Chapter 9

Fractions

Have you ever thought about how many times you use fractions every day? Whenever you share something with a friend, you are using fractions in some way. If you have ten pencils in your pencil box and a friend asks to borrow one, you have loaned $\frac{1}{10}$ of your pencils. If you shared a cookie with your friend by breaking it in two equal pieces, you have given $\frac{1}{2}$ of your cookie away.

Look at the fraction below.

This is probably the most common fraction we use and most likely the first fraction that came your way. The 1 in this fraction is called the *numerator*. The numerator is the number above the line and stands for the number of equal parts. The 2 in this fraction is called the *denominator* and is the total number of equal parts in the whole or in a set. The denominator is the number below the line.

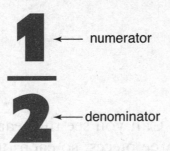

EQUAL PARTS OF A WHOLE

The most important thing to remember about fractions is that all the parts of the whole are *equal parts*. Look at the towers drawn for you below.

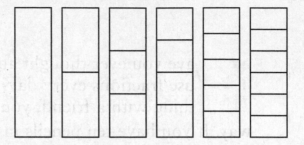

Each of these towers is the same size, but do you see that each of the pieces in each tower is the same? Let's look at just the tower that is divided into 4 pieces.

Each piece is the same size and has the same name. Since there are four parts, each piece is called one-fourth. It is written like this: $\frac{1}{4}$.

Now let's look at another tower that is divided into 3 pieces.

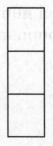

Can you see that each piece is the same size? There are three pieces, so each one is named one-third and is written like this: $\frac{1}{3}$.

How about the tower that is divided into two pieces? Do you know what each piece is named?

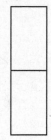

If you said one-half, you were right. It is written like this: $\frac{1}{2}$.

The tower we haven't looked at yet is the tower with five equal pieces.

Do all the pieces look the same? To be sure you can shade one of them in with your pencil. Since there are five equal pieces, the one you shaded is called one-fifth and is written like this: $\frac{1}{5}$.

EQUAL PARTS OF A GROUP

Suppose you had a group of things like the pencils in your pencil box we talked about before. The number of things you have altogether is the number in the denominator. In this case you have 10 pencils in your pencil box, so the denominator is ten. If you take the pencil back from your friend after it has been used and put it on the desk, the fraction of pencils on your desk is $\frac{1}{10}$. You can read this as one-tenth or you can say one out of ten. What is the fraction of pencils still in your pencil box? If you said $\frac{9}{10}$, you were right.

It is lunchtime and you see a table with some pieces of fruit on it. The picture below is what you see.

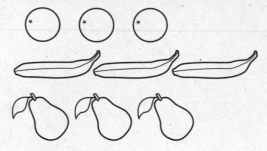

If you ate one banana, what fraction of the fruit did you eat?

The answer is $\frac{1}{8}$ because there are eight pieces of fruit and you ate one out of 8.

Does it matter what piece of fruit you ate?

No. As long as you ate just one of the eight pieces of fruit on the table, the fraction remains one-eighth.

If you ate all the oranges, what is the fraction of fruit that you ate?

The answer is $\frac{3}{8}$ because there are three oranges and there were eight pieces altogether.

TEST YOUR SKILLS—EQUAL PARTS OF A WHOLE AND EQUAL PARTS OF A GROUP

Circle the letter next to the best answer for each question.

1. Look at the picture of the rectangle below.

Which of the following is the fraction for the shaded part of this rectangle?

A. $\frac{1}{2}$

B. $\frac{1}{4}$

C. $\frac{1}{5}$

D. $\frac{1}{6}$

2. Study the circle drawn below.

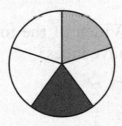

Which of the following is the fraction for the shaded parts?

A. $\frac{1}{5}$

B. $\frac{2}{5}$

C. $\frac{2}{3}$

D. $\frac{2}{4}$

3. Look at the picture of the shapes below.

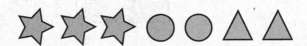

Which of the following is the fraction of stars in this group of shapes?

A. $\frac{1}{3}$

B. $\frac{3}{4}$

C. $\frac{3}{7}$

D. $\frac{1}{7}$

4. Look at the picture of shoes below.

Which of the following is the fraction of shoes that have laces?

A. $\frac{4}{6}$

B. $\frac{2}{4}$

C. $\frac{2}{6}$

D. $\frac{6}{4}$

5. Look at the picture of the crayons below.

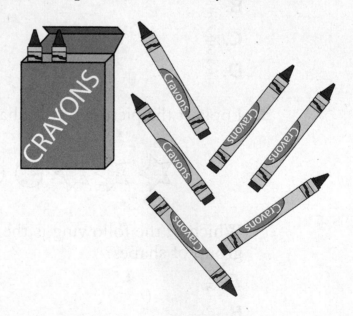

Which of the following is the fraction of crayons that are out of the box?

A. $\frac{2}{6}$

B. $\frac{2}{8}$

C. $\frac{6}{6}$

D. $\frac{6}{8}$

Answers on page 172.

UNIT FRACTIONS

Any fraction with a one as numerator is called a *unit fraction*. Actually, these fractions were the ones most commonly used by the ancient Egyptians.

Let's look at the towers of fractions we studied earlier in this chapter.

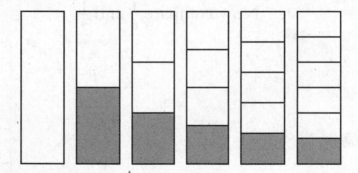

Do you see what happens when each tower has one part shaded? Did you notice that the greater the denominator the smaller the pieces? Now let's look at just the shaded parts of each tower.

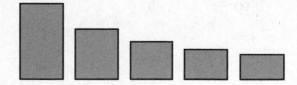

Does this look like a staircase to you? Every time the denominator gets bigger the piece gets smaller. Let's try comparing some of these unit fractions. Remember that < means less than and > means greater than. It might help to see the sideways L in the less than sign.

How does $\frac{1}{2}$ compare with $\frac{1}{5}$?

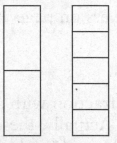

You can see that $\frac{1}{2}$ is clearly greater than $\frac{1}{5}$. This is how it is written: $\frac{1}{2} > \frac{1}{5}$.

Now compare $\frac{1}{6}$ and $\frac{1}{4}$.

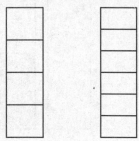

Can you see that $\frac{1}{6}$ is less than $\frac{1}{4}$? It is written this way: $\frac{1}{6} < \frac{1}{4}$.

What do you think would happen if the tower were divided into tenths? How do you think $\frac{1}{10}$ would compare to $\frac{1}{4}$?

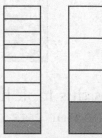

It's easy to see that $\frac{1}{10}$ is a lot less than $\frac{1}{4}$. It is written this way: $\frac{1}{10} < \frac{1}{4}$.

FRACTIONS ON A NUMBER LINE

If you are having difficulty comparing and ordering fractions, then you might find this activity helpful. A number line is something you have most likely been using since you learned how to count. In order to place fractions on a number line, it has to be expanded or magnified.

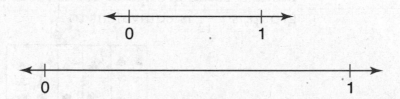

There is a lot of room on this new number line, and we will find a place for our fractions on it.

Let's begin with $\frac{1}{2}$. This fraction should be placed in the exact middle of this number line.

How about $\frac{1}{4}$? This fraction is placed exactly in the middle between 0 and $\frac{1}{2}$. If $\frac{1}{4}$ is halfway to $\frac{1}{2}$, then $\frac{2}{4}$ is the same as $\frac{1}{2}$, so put $\frac{2}{4}$ below $\frac{1}{2}$.

Now there's $\frac{1}{3}$ to deal with. This fraction is more than $\frac{1}{4}$ but less than $\frac{1}{2}$, so we will place it between $\frac{1}{4}$ and $\frac{1}{2}$ on the number line.

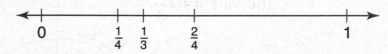

EQUIVALENT FRACTIONS

Fractions that name the same amount but use different numbers in the numerator and denominator are called *equivalent fractions*.

Here is how it works. You have six marbles in a bag and decide to share $\frac{1}{2}$ of them with a friend. You give your friend 3 of those 6, or $\frac{3}{6}$ of the marbles. So $\frac{3}{6}$ is equivalent or the same as $\frac{1}{2}$.

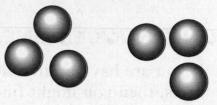

Suppose you have 12 playing cards and you need to use $\frac{1}{3}$ of them for a game. Spread them out on the table and make 3 equal groups of cards. There are 4 cards in each group, so $\frac{4}{12}$ is equivalent to $\frac{1}{3}$.

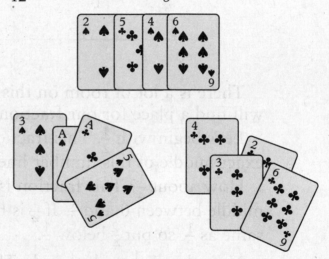

A pizza is cut into 8 slices. Your mother says you can eat $\frac{1}{4}$ of the pizza, so you take 2 of the 8 slices or $\frac{2}{8}$. So $\frac{2}{8}$ is equivalent to $\frac{1}{4}$.

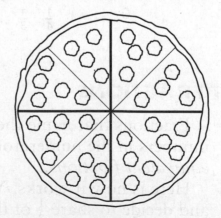

TEST YOUR SKILLS—UNIT FRACTIONS, FRACTIONS ON A NUMBER LINE, AND EQUIVALENT FRACTIONS

Circle the letter next to the best answer for each question.

1. Which of the following is a unit fraction?

    A. $\frac{3}{4}$

    B. $\frac{2}{7}$

    C. $\frac{1}{4}$

    D. $\frac{2}{3}$

2. Which of the following fractions is greater than $\frac{1}{3}$?

    A. $\frac{1}{2}$

    B. $\frac{1}{4}$

    C. $\frac{1}{5}$

    D. $\frac{1}{6}$

3. Study the number line shown below.

    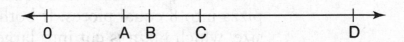

    Which of the following letters is in the right place for $\frac{1}{3}$?

    A. A

    B. B

    C. C

    D. D

4. Which of the following is equivalent to $\frac{1}{2}$?

    A. $\frac{3}{8}$

    B. $\frac{5}{7}$

    C. $\frac{3}{4}$

    D. $\frac{4}{8}$

5. Which of the following is a true math sentence?

    A. $\frac{1}{3} > \frac{1}{2}$

    B. $\frac{1}{2} > \frac{1}{8}$

    C. $\frac{1}{5} < \frac{1}{6}$

    D. $\frac{1}{10} > \frac{1}{4}$

Answers on pages 172–173.

PROBLEM SOLVING

Write your answers on the lines below each question.

1. Frank cut one pizza into 6 equal pieces and another pizza into 8 equal pieces. If both pizzas are the same size, which pizza is cut into larger pieces?

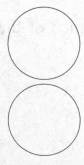

_____

_____

2. There are 7 days in a week. What fraction of the week is Monday?

_____

_____

3. What fraction of the months of the year begins with the letter M?

_____

_____

4. Study the fraction strips below. Which is greater: $\frac{1}{2}$ or $\frac{2}{3}$? Shade the strips in to prove your answer.

$\frac{1}{2}$	$\frac{1}{3}$
	$\frac{1}{3}$
$\frac{1}{2}$	$\frac{1}{3}$

_____

_____

5. Place the following fractions correctly on the number line below.

$$\frac{1}{2}, \frac{1}{4}, \frac{3}{4}, \frac{1}{3}, \frac{2}{3}$$

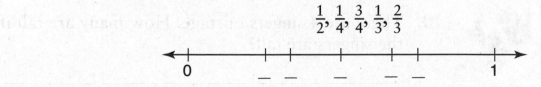

_____

_____

6. Morgan has 10 potato chips in her snack bag and Jane has 14 pretzels in her bag. If each of them decides to share $\frac{1}{2}$ of her snack, then why is $\frac{1}{2}$ of Morgan's chips equal to 5, and $\frac{1}{2}$ of Jane's pretzels equal to 7? Explain your answer on the lines below.

_____

_____

7. Ana's friends ate one-half of the cookies. If there were 18 cookies to begin with, how many cookies were left?

_____

_____

8. Susan has 6 bunnies. How many bunnies are white if only $\frac{1}{3}$ of Susan's bunnies are white?

_____

_____

9. Bob has 12 pencils in his desk. If $\frac{1}{4}$ of them are sharpened, how many pencils in Bob's desk are sharpened?

_____

_____

10. There are 4 singers onstage. How many are tall if $\frac{1}{2}$ of the singers are tall?

_____

_____

Answers on page 173.

Chapter
10

Problem Solving

Problem solving is just like anything else; once you practice it and learn some helpful hints, you can do it well. We will practice nine of the most common strategies in this chapter. If you work on these, you will be in good shape!

NUMERIC AND GEOMETRIC PATTERNS

Solving or reading patterns is something you have most likely been doing since you were quite young. Bead necklaces from kindergarten days may come to mind.

Geometric patterns are patterns that are made up of geometric shapes such as squares and circles. Study the pattern below and read it out loud.

◇ □ ○ ○ ○ ◇ □ ○ ○ ◇ ___, ___, ___.

What three shapes go in the blanks to complete the pattern?

◇ □ ○ ○ ○ ◇ □ ○ ○ ◇ □, ○, ○.

Let's try another one.

◇ ★ ● ✳ ◇ ★ ● ✳ ◇ ★ ● ✳ ◇ ★ ___, ___, ___.

145

Which shapes belong in the blanks?

The last shape before a blank is ★, so the next shape after that is ⬠. After that comes ✳, and the final shape is ◇.

Now let's explore numeric patterns. As the name suggests, these are patterns of numbers. Study the numeric pattern below.

$$1, 3, 5, 7, 9, 11, \_\_, \_\_\_, \_\_$$

How is this pattern formed? These are odd numbers beginning with 1 and we are counting by 2s. So 13, 15, and 17 complete the pattern.

Try this one.

$$3, 6, 9, 12, 15, 18, 21, 24, \_\_\_, \_\_\_, \_\_$$

This time the numbers are counting by 3s beginning with 3. The numbers that finish the pattern are 27, 30, and 33.

This is a bit more challenging.

$$46, 42, 38, \_\_\_, 30, \_\_\_\_, 22, 18$$

This time the numbers are counting backward. Did you notice that the difference between each number is 4? The two missing numbers are 34 and 26.

MAKE A TABLE

This strategy helps you organize your thoughts and information in order to solve the problem.

Jan is making necklaces for her friends. Every necklace has 12 beads. If Jan makes 6 necklaces, how many beads will she need?

Set up a table like the one below to help you solve the problem.

Necklaces	1	2	3	4	5	6
Beads	12	24	36	48	60	72

By setting up the table, you can see that the number of beads increases by 12 with each necklace. By the time you are at 6 necklaces you see that you need 72 beads. That's a lot of beads!

Let's try another one like that.

Ona loves shoes! She has 16 pairs in her closet. How many shoes does she have altogether?

Draw a table for all of Ona's shoes.

Pairs 1	2	3	4	5	6	7	8	9	10	11	12	13	14	15	16
Shoes 2	4	6	8	10	12	14	16	18	20	22	24	26	28	30	32

She has 32 shoes and most likely not much floor space in her closet!

FIND A PATTERN

In order to find a pattern it is sometimes necessary to write down all the information in a problem to see the pattern begin to form.

Mary is watching the flowers in her garden bloom. On the first day she sees one bloom. On the second day she sees 4 **more** blooms, making 5 flowers in bloom. On the third day she sees 4 **more** flowers bloom, making 9 flowers in bloom. How many flowers will be in bloom by the sixth day?

The pattern is that the number of flowers increases by 4 each day. Number sentences can be used to help you solve this problem.

1 flower on day 1.
1 + 4 = 5 on day 2.
5 + 4 = 9 on day 3.
9 + 4 = 13 on day 4.
13 + 4 = 17 on day 5.
17 + 4 = 21 on day 6.

This problem can also be solved using a table, but it may be helpful to write out the equations to keep you from making a mistake.

Nancy's class is doing a line dance. Every third dancer in the line will wear blue. The rest of them will wear black. Nancy is the 13th dancer in the line. What color will she wear?

Let's look for the pattern of blue. Every third dancer in line will wear blue, so we will make every third number blue.

1, 2, 3, 4, 5, 6, 7, 8, 9, 10, 11, 12, 13, 14, 15

Since 13 is not a blue number, Nancy will be wearing black.

WORKING BACKWARD

The trick with this strategy is to read the problem carefully to the very end. Try this one!

Brenda has a bag full of soup cans. Pat gives her 5 more cans, and Brenda puts 8 cans on the counter. There are 10 cans left in the bag. How many cans were in the bag to start?

You can start to solve this problem by using the last piece of information given to you. There are 10 cans still in the bag. There are also 8 cans on the counter, but Pat gave 5 of those to Brenda. Subtract the 5 cans Pat gave from the 8 cans on the counter.

$$8 - 5 = 3$$

Add those 3 cans to the 10 cans still in the bag.

$$3 + 10 = 13$$

The answer to the problem is that Brenda had 13 cans in the bag to start.

This strategy is often used in time problems.

Linnie wants to be on time for the movie. It takes her 15 minutes to walk to the theater and 15 minutes to eat lunch. What time should Linnie start eating lunch in order to get to the movies at 2 o'clock?

This problem asks us to go back in time for Linnie to eat her lunch and walk to the movies. Each activity takes 15 minutes, and the two together take 30 minutes because $15 + 15 = 30$. Another way to think of 30 minutes is one half-hour. One half-hour before 2 o'clock is 1:30, so Linnie must start eating her lunch at 1:30 in order to be on time for a 2 o'clock movie.

USING LOGICAL REASONING

This strategy requires a lot of patience because you have to give some thought to your answer.

Here is a quick example.

George eats breakfast at 6 o'clock. He also goes to a play at 6 o'clock. How is this possible?

The answer is that George eats breakfast at 6 A.M. and goes to the play at 6 P.M.

This is a more common example.

Kelly is arranging her sweaters on a shelf. The pink sweater is on the right. The yellow sweater is not next to the pink sweater. The red sweater is not next to the pink sweater, and the blue sweater is next to the yellow sweater.

To solve a problem like this it is best to get four small pieces of paper and write the names of the colors on each piece. Now move the pieces around until you have the correct order.

Red, yellow, blue, pink is the answer.

GUESS AND CHECK

Your parents most likely know this strategy as trial and error. The meaning is the same, but guess and check sounds friendlier! For this strategy you try out an answer and check to see if it works. If it doesn't work, check to see if it is too high or too low and then try another number.

Here is a good example.

The sum of my digits is 11. The product of my digits is 28. I am less than 70. What number am I?

The first step to solve this problem would be to think of two numbers that add to 11. There are a few.

$$10 + 1, 9 + 2, 8 + 3, 7 + 4, \text{ and } 6 + 5$$

Now look at these pairs and decide which pair of numbers has a product of 28.

$$7 \times 4 = 28$$

Put them together and you have 74. Is 74 the answer?

Check again and reverse the order so the number is less than 70.

The answer is 47.

Try this one with the guess-and-check method.

The Haring family is going to see a play. They bought some children's tickets for $4 each and some adult tickets for $8 each. They bought one more children's ticket than adult ticket. They spent $28. How many children's and adult tickets did they buy?

To solve the problem, start by trying 2 children's tickets for $8 and one adult ticket for $8 for a total of $16. That's too little. Try 4 children's tickets for $16 and 3 adult tickets for $24 for a total of $40. That's too much. Now try 3 children's tickets for $12 and 2 adult tickets for $16 for a total of $28!

ACT IT OUT

This strategy is fun because you can try it yourself! Most often, this strategy can help you solve a problem related to probability.

There are six marbles in a bag. Four of them are black, and two of them are white. Since the number of black marbles is twice the number of white marbles, the probability of pulling out a black marble should be twice as good. To prove it, get out a bag and place in it any objects that are the same size and shape. Be sure that there are six in the bag and that the number of objects of one color is twice the number of objects of the other color. Pull out one object at a time without looking and keep track of your results. The more you do this activity, the closer you will come to the correct probability of pulling out a black marble twice as often as a white marble.

You might also use this strategy when solving some logical reasoning problems.

Here is one example.

There are 10 people standing in line to get into the movies. Karen is standing two people in front of the person who is second from the end of the line. Where is Karen standing in line?

Ask ten people to stand in line and decide who is standing in the front of the line. Then find the person who is standing second from the end of the line and move two people forward from there to find Karen. She is sixth in line.

SOLVE A SIMPLER PROBLEM

Sometimes you need to use simpler numbers in a problem to help you decide what operation to use. Then you can go back to the larger numbers and find the exact answer.

Let's try this one.

Central High School is having a spring musical next week. The students have sold 2,853 tickets for the first night and 4,521 for the second night. How many tickets have they sold for both nights?

These are big numbers, so let's make things simpler by thinking of 20 tickets sold for the first night and 40 tickets sold for the second night. In order to find how many tickets were sold for both nights, we add 20 and 40, for 60 tickets. That makes sense because the number of tickets sold for both nights should be bigger than the number of tickets sold for just one night. Now go back to the problem and add 2,853 + 4,521, for a total of 7,374 tickets.

This strategy can work with other operations as well. Try this one.

Alyssa is taking her two children and their friend to the amusement park. The cost for admission is $4.95 for each child and $6.95 for each adult. How much will it cost to get into the park?

Make the problem simpler by rounding $4.95 to $5 and $6.95 to $7. It will cost 3 × $5 for the three children and $7 for Alyssa. Since 3 × $5 = $15 and $15 + $7 = $22, then this is a good start toward solving this problem with the exact numbers.

The exact answers are 3 × $4.95 = $14.85.

$$\$14.85 + \$6.95 = \$21.80$$

This answer is very close to the one we got using simpler numbers.

DRAW A PICTURE

This strategy might be the most fun of all. It is a natural thing to want to draw a picture to help you solve some problems.

Let's try this one.

Peter has just planted his spring garden of lettuce and radishes. He wants to keep the rabbits out, so he will put a fence around it. The garden is rectangular and is 8 feet long and 5 feet wide. How long a fence will Peter need to go around his garden?

Draw a rectangle and label the long side 8 feet and the short side 5 feet. Remember that the other long side is also 8 feet and the other short side is also 5 feet. Add these four numbers, 8 + 8 + 5 + 5, to get 26 feet. Peter needs to buy 26 feet of fence to go around his lettuce and radish garden.

Look at the picture drawn for you below to see if your picture is the same.

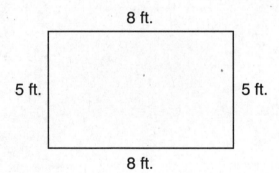

A picture can also be drawn for a problem like this one.

Autumn is waiting in line to buy a ticket for the movies. She notices that there are 11 people in front of her and 6 people behind her. How many people are in line for the movies?

Draw the ticket window for the front of the line and draw stick people standing in line. Draw 11 people and then draw Autumn. Now draw 6 more people. Add 11 + 1 + 6 to get a total of 18 people. Be careful to count

Autumn. That is the only trick in this problem! Study the picture below to see if you drew the right picture.

ANSWERS TO PRACTICE PROBLEMS

CHAPTER 1: PLACE VALUE

TEST YOUR SKILLS SOLUTIONS—READ AND WRITE NUMBERS TO 1,000 AND USE EXPANDED FORM TO 999 (PAGES 5-6)

1. C. The answer is eight hundred thirty-one.

2. D. The answer is 759.

3. B. The answer is 682.

4. A. The answer is 900 + 60 + 3.

5. C. The answer is four hundred fifty.

TEST YOUR SKILLS SOLUTIONS—COMPARE AND ORDER NUMBERS TO 1,000 AND PLACE NUMBERS ON A NUMBER LINE (PAGES 9-10)

1. 34 $<$ 43

2. 150 $>$ 105

3. 998 $<$ 999

4. D. 618, 625, 629, 631

5. B. 956, 901, 899, 856

6. The numbers are placed on the number line as shown below:

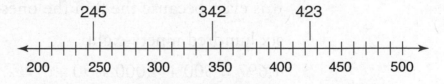

© Copyright 2012 by Barron's Educational Series, Inc.

TEST YOUR SKILLS SOLUTIONS—IDENTIFY ODD AND EVEN NUMBERS AND SKIP COUNT BY 25s, 50s, AND 100s TO 1,000 (PAGES 13-15)

1. **B.** 85 is odd because of the 5 in the ones place.

2. **C.** 78 is even because of the 8 in the ones place.

3. **C.** 481 is odd because of the 1 in the ones place.

4. **B.** 425 is missing between 400 and 450.

5. **D.** 1,000 is the next number in the series because 950 + 50 = 1,000.

6. **C.** 75 is the answer because when skip counting by 25, 75 comes after 50.

PLACE VALUE PUZZLES SOLUTIONS (PAGES 16-17)

For each solution below, check to see if each clue is answered correctly.

1. 543

2. 157

3. 969

4. 654

5. 4,790

PROBLEM-SOLVING SOLUTIONS (PAGES 17-18)

1. 500 + 0 + 9 = 509

2. 750, 775. This pattern skip counts by 25.

3. It is even because the 8 in the ones place is even.

4. six hundred ninety-eight

5. 4,697 = 600 + 4,000 + 90 + 7

6. 721, 712, 217, 271, 172, 127. List all the numbers with 7 in the hundreds place, then 2 in the hundreds place, and then 1 in the hundreds place.

7. 981, 918, 891, 819. Look to the hundreds place for the greatest number, then go to the tens place and finally the ones place.

8. 781 ≤ 871

9. 532 or 352. Either of these is the answer because you must have the 2 in the ones place to make it an even number.

10. A 3-digit number is always greater than a 2-digit number because the greatest 2-digit number is 99, and that is less than 100 or anything greater than that.

CHAPTER 2: ADDITION

TEST YOUR SKILLS SOLUTIONS—COMMUTATIVE PROPERTY, IDENTITY PROPERTY, AND ASSOCIATIVE PROPERTY (PAGES 24-25)

1. **B.** commutative—The order of addends does not change the sum, so 8 would correctly solve the problem.

2. **C.** identity—The only number that can be added to 16 to result in a sum of 16 is 0, so the identity property is the answer.

3. **A.** associative—Look for parentheses to tell you this is the associative property.

4. **A.** 5—Using the commutative property, the only solution is 5.

5. **D.** 32—Using the identity property, 32 is the only correct choice.

© Copyright 2012 by Barron's Educational Series, Inc.

TEST YOUR SKILLS SOLUTIONS—PATTERNS IN ADDITION AND ODD AND EVEN SUMS (PAGES 29-30)

1. **D.** 10,000. Add the basic fact $5 + 5 = 10$ and add 3 zeroes to make 10,000.

2. **C.** 1,000. Add the basic fact $6 + 4 = 10$ and add 2 zeroes.

3. **B.** 80. Add the basic fact $9 + 8 = 17$ and add 1 zero.

4. **B.** $5 + 7$. Two odd addends are needed to make an even sum.

5. **C.** $7 + 2 = 9$. To make an odd sum there must be one even addend and one odd addend.

TEST YOUR SKILLS SOLUTIONS—MORE THAN ONE WAY TO ADD AND ADDING 3-DIGIT NUMBERS (PAGE 33)

1.
$$\begin{array}{r} 1 \\ 683 \\ +925 \\ \hline 1608 \end{array}$$

2.
$$\begin{array}{r} 1 \\ 719 \\ +456 \\ \hline 1175 \end{array}$$

3.
$$\begin{array}{r} 1 \\ 837 \\ +939 \\ \hline 1776 \end{array}$$

4.
$$\begin{array}{r} 1\,1 \\ 269 \\ +874 \\ \hline 1143 \end{array}$$

5.
$$\begin{array}{r} \overset{1}{378} \\ +609 \\ \hline 987 \end{array}$$

TEST YOUR SKILLS SOLUTIONS—ESTIMATING SUMS (PAGE 35)

1. 58 rounded to the nearest 10 is 60.

2. 61 rounded to the nearest 10 is 60.

3. 334 rounded to the nearest 100 is 300.

4. 928 rounded to the nearest 100 is 900

5. 251 rounded to the nearest 100 is 300.

TEST YOUR SKILLS SOLUTIONS—ADDING COINS AND BILLS USING $0.00 TO INDICATE SUMS (PAGE 37)

For this activity it is important to have the decimal points lined up.

1. $38.90

2. $26.95 (Remember, $25 can be written $25.00 in order to add it to $1.95.)

3. $31.00 or $31

4. $30.00 or $30

5. $19.72

PROBLEM-SOLVING SOLUTIONS (PAGES 38-40)

1. D. Look for the parentheses!

2. B. 5 is odd and 2 is even and 8 is even.

3. C. 25¢ + 5 ¢ + 5¢ + 1¢ = 36¢

4. A. 36 has an even number (6) in the ones place.

© Copyright 2012 by Barron's Educational Series, Inc.
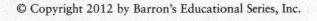

5. **B.** This number rounded to the nearest hundred is
500 + 200 = 700.

6. **C.** When you see a zero as an addend, you have the
identity property.

7. **B.** 6 + 4 = 10, and one more zero makes 100.

8. **D.** 2 is even, 8 is even, and 10 is even.

9. **B.** The order of the addends is different, but the sum
is the same.

10. **C.** 72 rounded to the nearest 10 is 70, so put a 3 in
the hundreds place and you have it!

EXTENDED-RESPONSE QUESTIONS SOLUTIONS (PAGE 41)

A. The explanation should include rounding to the nearest whole dollar, then adding the rounded dollars.
$3.95 rounds to $4, $4.50 rounds to $5, and $8.95
rounds to $9. The estimated sum is $18.00.

B. The exact sum is $3.95 + $4.50 + $8.95 = $17.40.

C. $20.00 − $17.40 = $2.60

$$
\begin{array}{r}
{\scriptstyle 1\ 9\ 10} \\
\$\cancel{2\,0}.\cancel{0}0 \\
-17.40 \\
\hline
\$2.60
\end{array}
$$

CHAPTER 3: SUBTRACTION

TEST YOUR SKILLS SOLUTIONS—FACT FAMILIES AND PATTERNS IN SUBTRACTION (PAGES 47-48)

1. **A.** 3 + 4 = 7. This is the only answer choice with 7,
3, and 4.

2. D. 6 + 8 = 14. This is the only answer choice with 8, 6, and 14.

3. B. 8 + 7 = 15. This is the only answer choice with 7, 8, and 15.

4. D. <u>12</u>0 – <u>4</u>0 = <u>8</u>0. The basic fact has been underlined.

5. B. 20 – 16 = 4. If you were to add a zero to make a new ones place in each number you would have the original number sentence.

TEST YOUR SKILLS SOLUTIONS—REGROUPING IN SUBTRACTION AND REGROUPING ACROSS ZEROES (PAGES 52-53)

1.
$$\begin{array}{r} {}^{7}\,{}^{11} \\ \cancel{8}\cancel{1} \\ -53 \\ \hline 28 \end{array}$$

2.
$$\begin{array}{r} {}^{5}\,{}^{13} \\ \cancel{6}\cancel{3} \\ -35 \\ \hline 28 \end{array}$$

3.
$$\begin{array}{r} {}^{4}\,{}^{18} \\ \cancel{5}\cancel{8} \\ -29 \\ \hline 29 \end{array}$$

4.
$$\begin{array}{r} {}^{9} \\ {}^{2}\,\cancel{10}\,10 \\ \cancel{3}\,\cancel{0}\,\cancel{0} \\ -136 \\ \hline 164 \end{array}$$

5.
$$\begin{array}{r} {}^{9} \\ {}^{5}\,\cancel{10}\,10 \\ \cancel{6}\,\cancel{0}\,\cancel{0} \\ -374 \\ \hline 226 \end{array}$$

© Copyright 2012 by Barron's Educational Series, Inc.

TEST YOUR SKILLS—ESTIMATING DIFFERENCES, ODD OR EVEN DIFFERENCES, AND SUBTRACTING 3-DIGIT NUMBERS (PAGES 56-57)

1. **C.** $11 - 5 = 6$. This is the only odd minus odd problem.

2. **C.** 66 rounds to 70, and 47 rounds to 50, so $70 - 50 = 20$.

3. **A.**

$$
\begin{array}{r}
\overset{4\ 16}{7\cancel{5}\cancel{6}} \\
-437 \\
\hline
319
\end{array}
$$

4. **D.** $8 - 3 = 5$

5. **D.** 89 rounds to 90, and 32 rounds to 30, so $90 - 30 = 60$.

PROBLEM-SOLVING (PAGES 58-59)

1. This problem requires rounding. 863 rounds to 900, and 345 rounds to 300, so $900 - 300 = 600$.

2. $856 - 598 = 258$ cards

3. $145 - 105 = 40$ minutes

4. $9 - 5 = 4$ years

5. $61 - 23 = 38$ students

6. $100 - 37 = 63$ cards

7. $900 - 176 = 724$ videos

8. This problem requires rounding for an estimate. 387 rounds to 400, and 165 rounds to 200, so $400 - 200 = 200$ books.

9. $56 - 27 = 29$ cars

10. $46 - 29 = 17$ boys

EXTENDED-RESPONSE QUESTIONS SOLUTIONS (PAGE 59)

1. The 60 minutes need to be separated into three groups of the same amount. 20 + 20 + 20 = 60, so Shane practiced 20 minutes and Ben practiced 40 minutes.

2. 15 – 6 = 9 people on the bus at the second stop. 9 + 8 = 17 people on the bus at the third stop.

3. $1.45 + $6.55 = $8 and $10 – $8 = $2.

CHAPTER 4: TELLING TIME

TEST YOUR SKILLS SOLUTIONS—TELLING TIME TO THE HOUR AND TELLING TIME TO THE HALF-HOUR (PAGES 63-64)

1. **D.** The hour hand is halfway between 8 and 9, and the minute hand is on the 6.

2. **A.** The hour hand is halfway between the 6 and 7, and the minute hand is on the 6.

3. **C.** The hour hand is halfway between the 12 and the 1, and the minute hand is on the 6.

TEST YOUR SKILLS SOLUTIONS—TELLING TIME TO THE MINUTE (PAGES 67-68)

1. **A.** The hour hand is just about on the 6, and the minute hand is on the 10.

2. **D.** The hour hand is between the 3 and the 4, but closer to the 4, and the minute hand is on the 9.

3. **B.** The time is 9:15 because the hour hand is just past the 9 and the minute hand is on the 3.

4. **A.** The time is 7:50 because the hour hand is just before the 8 and the minute hand is on the 10.

© Copyright 2012 by Barron's Educational Series, Inc.

PROBLEM-SOLVING SOLUTIONS (PAGE 69)

1. 30 minutes is 2 quarter-hours because there are 15 minutes in one quarter-hour and 15 + 15 = 30.

2. **D.** 12:15 P.M. is just after 12 noon.

3. **C.** The hour hand is slightly past the 1, and the minute hand is on the 3.

4. quarter to eight

5. The hour hand is halfway between the 8 and the 9.

EXTENDED-RESPONSE QUESTIONS SOLUTIONS (PAGE 70)

1. 9:15 P.M. or a quarter past 9

2. 1:45 P.M. or a quarter to 2

3. This is closer to 4:00 P.M. because it is 25 minutes before 4 and 35 minutes after 3.

CHAPTER 5: DATA AND GRAPHS

TEST YOUR SKILLS SOLUTIONS—QUESTIONS FOR FAVORITE FRUITS IN BRADLEY'S CLASS (PAGE 76)

1. Since the bar over apples is the tallest, Bradley's class liked apples the most.

2. There were 6 students who liked oranges and 4 students who liked bananas. Since 6 – 4 = 2, then 2 students liked oranges better than bananas.

3. Since the bar over bananas is the smallest, then the students liked bananas the least.

4. Since 10 students liked apples, 4 students liked bananas, and 5 students liked grapes, and since 4 + 5 = 9 and 10 – 9 =1, then 1 student liked apples more than bananas and grapes.

5. There are 10 students who liked apples. Since
6 + 4 = 10, then oranges and bananas were the two
fruits they liked as much as they liked apples.

PROBLEM-SOLVING SOLUTIONS (PAGES 76–77)

MS. HORNUNG'S MUFFINS

Type of Muffin	Number of Muffins
Blueberry	○○○○○○○
Corn	○○○○○
Orange	○○○
Cranberry	○○○○

Key: ○ = 2 muffins

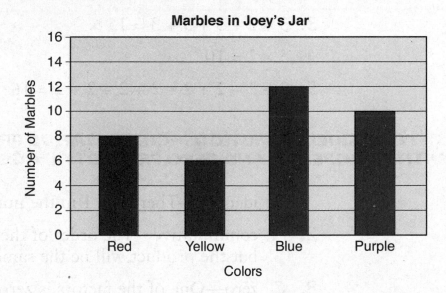

CHAPTER 6: MULTIPLICATION AND DIVISION

TEST YOUR SKILLS SOLUTIONS—MULTIPLICATION IS REPEATED ADDITION, MULTIPLICATION AS AN ARRAY, AND TWO DIFFERENT PROBLEMS WITH THE SAME PRODUCT (PAGES 82-83)

1. The product is 18.
2.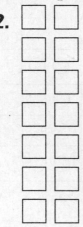
3. $3 + 3 + 3 + 3 + 3 = 15$
4. $5 \times 2 = 10$
5. $2 + 2 + 2 + 2 + 2 + 2 + 2 + 2 = 16$

TEST YOUR SKILLS SOLUTIONS—COMMUTATIVE PROPERTY, IDENTITY PROPERTY, AND ZERO PROPERTY (PAGES 86-87)

1. **B.** identity—There is a 1 in the number sentence.
2. **A.** commutative—The order of the factors is different but the product will be the same.
3. **C.** zero—One of the factors is zero, so the product is zero.
4. **A.** 7—The factors need to be the same on both sides of the equal sign.
5. **B.** 89—The identity property states that when 1 is a factor the product will be the other factor.

PROBLEM-SOLVING SOLUTIONS (PAGE 91)

1. $15 \div 3 = 5$ boxes

2. $3 \times 5\text{¢} = 15\text{¢}$

3. $6 \times 4\text{¢} = 24\text{¢}$

4. $5 \times 4 = 20$ T-shirts

5. $24 \div 3 = 8$ cans of tennis balls

EXTENDED-RESPONSE SOLUTIONS (PAGES 91–92)

1. $3 \times 5 = 15$ and $5 \times 3 = 15$, so both Gawain and Spencer had the same amount of pictures in their albums.

2. $3 \times 2 = 6$ and $4 \times 3 = 12$ $12 + 6 = 18$

18 people could ride in the canoes and rowboats altogether.

3. $3 \times 4 = 12$ and $2 \times 4 = 8$, so $12 + 8 = 20$ strings altogether.

4. $6 \times 4 = 24$ $30 - 24 = 6$

Cali needs 6 more photos to fill her album.

5. Here is a drawing showing the 4 rows of students.

$28 \div 4 = 7$ students in each row.

© Copyright 2012 by Barron's Educational Series, Inc.

CHAPTER 7: MEASUREMENT

TEST YOUR SKILLS SOLUTIONS—LINEAR MEASUREMENT (CUSTOMARY), WHOLE INCHES AND HALF-INCHES, FEET, AND YARDS (PAGES 96-97)

1. A. inches—In this case, 25 feet would be longer than the room, much less the desk, and yards would be even longer.

2. A. inches—Most pencils are 7 inches long.

3. B. feet—A person could not get into a car 10 inches long, and a car 10 yards long is longer than most houses.

4. C. yards—100 inches is way too small and there are 90 feet between each base, so 100 yards is the only one that will fit.

5. A. inches—A book that is 15 feet wide cannot be lifted, and a book 15 yards wide would not fit into a classroom.

TEST YOUR SKILLS SOLUTIONS—WEIGHT AND CAPACITY (PAGES 99-100)

1. A. laptop computer—The other items weigh very little.

2. A. inches—Computer keyboards are getting smaller rather than larger.

3. B. feet—No one can get into a bus 20 inches long, and 20 yards long is longer than most school buildings.

4. D. gallons—A bath tub holds a lot of water, and the gallon is the largest measure of water.

5. A. cups—A juice box is a small container, so the cup, which is the smallest measure, should be used.

PROBLEM-SOLVING SOLUTIONS (PAGE 101)

1. 3 + 3 = 6 pounds, so two boxes weigh 6 pounds.

2. The table and the desk are the same width because 1 yard = 3 feet.

3. 9 × 3 = 27 feet, so Sam's backyard is 27 feet wide.

4. This line is a little more than $2\frac{1}{2}$ inches long, so it is 3 inches to the nearest inch.

5. 2 gallons is more than 2 quarts because 1 gallon = 4 quarts, so Maria bought more milk.

EXTENDED-RESPONSE SOLUTIONS (PAGE 102)

1. Millie can use feet to measure the length of the wall and the length of the bookcase.

2. Jaden can use a ruler and draw a picture of the string on the ruler. If he carefully marks 4 inches as he goes, Jaden will see that there are exactly 3 pieces of string, and each will be 4 inches long.

3. If one pound is 16 ounces, then $\frac{1}{2}$ pound is 8 ounces. 16 + 8 = 24 ounces of candy.

CHAPTER 8: GEOMETRY

TEST YOUR SKILLS SOLUTIONS—THREE-DIMENSIONAL OR SOLID FIGURES (PAGES 109-110)

1. **C.** The word "angle" is not used to describe a solid figure.

2. **D.** If you can imagine a tree trunk without branches or leaves, you can imagine how much it looks like a cylinder.

© Copyright 2012 by Barron's Educational Series, Inc.

3. **B.** Look at the chart to check this answer.

4. **D.** Look at the chart to check this answer.

TEST YOUR SKILLS SOLUTIONS—POLYGONS (PAGES 114-115)

1. **D.** The triangle has only three sides.

2. **C.** The pentagon has five sides.

3. **A.** A circle is not a polygon because it has no line segments for sides.

4. **C.** The trapezoid has four sides, and only two of those sides are parallel.

TEST YOUR SKILLS SOLUTIONS—CONGRUENT AND SIMILAR FIGURES (PAGES 118-120)

1. **B.** Although they are turned a bit, these are exactly the same rhombus. So they are congruent.

2. **C.** These trapezoids are exactly the same, but one is upside-down.

3. **D.** These are both pentagons, but they are very different.

4. **C.** These circles are exactly the same.

TEST YOUR SKILLS SOLUTIONS—LINE OF SYMMETRY (PAGES 124-125)

1. **B.** The square has 4 lines of symmetry.

2. **B.** The rectangle has 2 lines of symmetry. One goes north and south, and the other goes east and west.

3. **D.** The tree has one line of symmetry running north and south.

4. **A.** The parallelogram has no lines of symmetry.

PROBLEM-SOLVING SOLUTIONS (PAGE 126)

1. A cube and a prism have the same number of faces, edges, and vertices.

2. A cube is different from a rectangular prism because all the faces are square in shape, and some or all the faces of a rectangular prism are rectangles.

3. A sphere has no edges or vertices because it has no faces.

4. Sam needed six different colors.

5. Jamisen now has a rectangular prism.

6. Cooper has a cone and a cylinder.

7. A square has four lines of symmetry because all four sides are equal. A rectangle has only two lines of symmetry because only the opposite sides are equal.

8. A circle has more lines of symmetry than you can draw. It has an infinite number of lines of symmetry.

9. Both a square and a rhombus have four equal sides.

10. A square always has four right angles, but a rhombus doesn't.

11. A parallelogram and a triangle are not similar because a parallelogram has four sides and a triangle has three sides.

12. All circles are similar because they are all formed the same way. The only difference is the size of the circles.

EXTENDED-RESPONSE QUESTIONS SOLUTIONS (PAGES 127–129)

1. The figures will be correctly identified. Two ways to identify a rectangle would include 4 sides, opposite sides equal, 4 right angles.

© Copyright 2012 by Barron's Educational Series, Inc.

2. The party hat is a cone. The basketball is a sphere. The can of soup is a cylinder. These shapes all have no edges.

3. The rectangle has 2 lines of symmetry. The square has 4 lines of symmetry. The heart has one line of symmetry.

4. All squares are similar because all squares have four equal sides and four square corners or right angles.

CHAPTER 9: FRACTIONS

TEST YOUR SKILLS SOLUTIONS—EQUAL PARTS OF A WHOLE AND EQUAL PARTS OF A GROUP (PAGES 134–137)

1. **D.** One section out of 6 is shaded, so the fraction is $\frac{1}{6}$.

2. **B.** Two sections out of 5 are shaded, so the fraction is $\frac{2}{5}$.

3. **C.** There are three stars out of seven shapes, so the fraction of stars is $\frac{3}{7}$.

4. **A.** Four out of six shoes have laces, so the fraction of shoes with laces is $\frac{4}{6}$.

5. **D.** There are eight crayons, and six of them are on the table, so the fraction is $\frac{6}{8}$.

TEST YOUR SKILLS SOLUTIONS—UNIT FRACTIONS, FRACTIONS ON A NUMBER LINE, AND EQUIVALENT FRACTIONS (PAGES 141–142)

1. **C.** $\frac{1}{4}$. This is a unit fraction because 1 is the numerator.

2. **A.** $\frac{1}{2}$. This fraction is the largest unit fraction there is.

3. **B.** $\frac{1}{3}$ is greater than $\frac{1}{4}$ but less than $\frac{1}{2}$.

4. **D.** $\frac{4}{8}$. The numerator is half of the denominator, and 4 pieces of a group of 8 is one-half of the group.

5. **B.** $\frac{1}{2} > \frac{1}{8}$. $\frac{1}{2}$ is greater than $\frac{1}{8}$ because $\frac{1}{2}$ is the greatest unit fraction and because there are only 2 pieces of the whole so the pieces are the biggest fraction pieces you can get.

PROBLEM-SOLVING SOLUTIONS (PAGES 142-144)

1. The six pizza slices are larger because there are fewer slices.

2. $\frac{1}{7}$ because there are 7 days in a week and Monday is just one of them.

3. $\frac{2}{12}$. The 2 months of the 12 that begin with M are March and May.

4. $\frac{2}{3} > \frac{1}{2}$. When $\frac{1}{2}$ is shaded and $\frac{2}{3}$ is shaded, the answer is shown.

5.
$$\begin{array}{cccccccc} 0 & & \frac{1}{4} & \frac{1}{3} & & \frac{1}{2} & \frac{2}{3} \ \frac{3}{4} & & 1 \end{array}$$

6. Jane had more pretzels to begin with, so her half is greater in number.

7. $\frac{1}{2}$ of 18 is 9 cookies.

8. $\frac{1}{3}$ of 6 is 2 bunnies.

9. $\frac{1}{4}$ of 12 is 3 pencils.

10. $\frac{1}{2}$ of 4 is 2 singers.

CHAPTER 10: PROBLEM SOLVING

Practice problems are found within the chapter.

© Copyright 2012 by Barron's Educational Series, Inc.

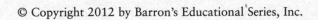

Sample Test 1

Each of the following two chapters has a practice test for you to take in your own home. After you finish each test, turn to the answer section to see how well you did. The only materials you will need are a pencil and a ruler. Remember to follow the directions in Part 2 especially. You will not get full credit for an answer that does not show your work if the question clearly asks for it!

When you take the test in school you will have 60 minutes to complete Part 1 and 40 minutes to complete Part 2. You might want to see how long it takes you to do each of these parts for practice. Very few students have had trouble finishing in the time allowed.

ANSWER SHEET: SAMPLE TEST 1

PART I

1. Ⓐ Ⓑ Ⓒ Ⓓ
2. Ⓐ Ⓑ Ⓒ Ⓓ
3. Ⓐ Ⓑ Ⓒ Ⓓ
4. Ⓐ Ⓑ Ⓒ Ⓓ
5. Ⓐ Ⓑ Ⓒ Ⓓ
6. Ⓐ Ⓑ Ⓒ Ⓓ
7. Ⓐ Ⓑ Ⓒ Ⓓ
8. Ⓐ Ⓑ Ⓒ Ⓓ
9. Ⓐ Ⓑ Ⓒ Ⓓ
10. Ⓐ Ⓑ Ⓒ Ⓓ
11. Ⓐ Ⓑ Ⓒ Ⓓ
12. Ⓐ Ⓑ Ⓒ Ⓓ
13. Ⓐ Ⓑ Ⓒ Ⓓ
14. Ⓐ Ⓑ Ⓒ Ⓓ

15. Ⓐ Ⓑ Ⓒ Ⓓ
16. Ⓐ Ⓑ Ⓒ Ⓓ
17. Ⓐ Ⓑ Ⓒ Ⓓ
18. Ⓐ Ⓑ Ⓒ Ⓓ
19. Ⓐ Ⓑ Ⓒ Ⓓ
20. Ⓐ Ⓑ Ⓒ Ⓓ
21. Ⓐ Ⓑ Ⓒ Ⓓ
22. Ⓐ Ⓑ Ⓒ Ⓓ
23. Ⓐ Ⓑ Ⓒ Ⓓ
24. Ⓐ Ⓑ Ⓒ Ⓓ
25. Ⓐ Ⓑ Ⓒ Ⓓ
26. Ⓐ Ⓑ Ⓒ Ⓓ
27. Ⓐ Ⓑ Ⓒ Ⓓ
28. Ⓐ Ⓑ Ⓒ Ⓓ

29. Ⓐ Ⓑ Ⓒ Ⓓ
30. Ⓐ Ⓑ Ⓒ Ⓓ
31. Ⓐ Ⓑ Ⓒ Ⓓ
32. Ⓐ Ⓑ Ⓒ Ⓓ
33. Ⓐ Ⓑ Ⓒ Ⓓ
34. Ⓐ Ⓑ Ⓒ Ⓓ
35. Ⓐ Ⓑ Ⓒ Ⓓ
36. Ⓐ Ⓑ Ⓒ Ⓓ
37. Ⓐ Ⓑ Ⓒ Ⓓ
38. Ⓐ Ⓑ Ⓒ Ⓓ
39. Ⓐ Ⓑ Ⓒ Ⓓ
40. Ⓐ Ⓑ Ⓒ Ⓓ

PART II

Use the space in the test booklet to show your work and your answers.

© Copyright 2012 by Barron's Educational Series, Inc.

SAMPLE TEST 1

PART 1

Answers on pages 199–202.

Directions: Choose the best answer for each question and fill in the correct bubble on your answer sheet.

---

1. Which of the following shirts has an even number on it?

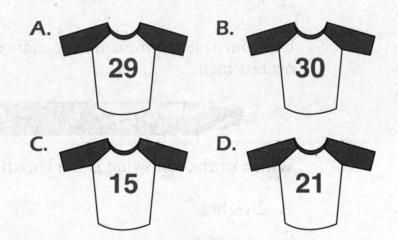

   A. 29   B. 30

   C. 15   D. 21

2. Which of the following is a pentagon?

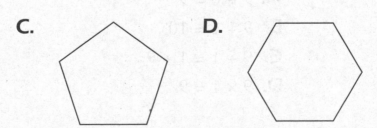

   A.   B.

   C.   D.

Go On

© Copyright 2012 by Barron's Educational Series, Inc.

3. Cassandra wrote the number sentence below but left the square without a number in it.

$$32 + 18 = \boxed{} + 32$$

Which of the following completes the number sentence correctly?

A. 18

B. 32

C. 50

D. 82

4. Use your ruler to measure the marker below to the nearest inch.

Which of the following is the length of the marker?

A. 2 inches

B. $2\frac{1}{2}$ inches

C. 3 inches

D. $3\frac{1}{2}$ inches

5. Which of the following number sentences is an example of the identity property for addition?

A. $9 + 0 = 9$

B. $9 + 1 = 10$

C. $9 + 1 = 1 + 9$

D. $9 \times 1 = 9$

6. Fran bought socks for $3.95. If she paid for them with a five-dollar bill, how much change did Fran get back?

 A. $8.95

 B. $2.95

 C. $1.95

 D. $1.05

7. What time does this clock show?

 A. 1:03

 B. 1:15

 C. 3:01

 D. 3:05

© Copyright 2012 by Barron's Educational Series, Inc.

Go On

8. Sue has four pets. Look at the picture. What fraction of Sue's pets are rabbits?

A. $\frac{1}{2}$

B. $\frac{1}{3}$

C. $\frac{1}{4}$

D. $\frac{3}{4}$

9. Saydee has 27 hard candies to share with her 9 friends. If each friend gets the same number of candies, how many hard candies will each friend receive?

A. 27

B. 18

C. 9

D. 3

10. Which of the following is pictured below?

    A. cone

    B. cube

    C. sphere

    D. pyramid

11. Which digit is in the tens place in the number shown below?

7,943

    A. 7

    B. 9

    C. 4

    D. 3

12. Which is the best way to measure the water in your bathtub?

    A. cups

    B. pints

    C. quarts

    D. gallons

Go On

© Copyright 2012 by Barron's Educational Series, Inc.

13. Lucas has a lot of books on his bookshelf. He has arranged them so he can find books easily. He made the graph below showing the topics of his books.

LUCAS'S BOOKS

Topic	Number of Books
Sports	 📖 📖 📖 📖 📖
Mystery	 📖 📖 📖
Travel	 📖
Biography	 📖 📖 📖 📖

Key: Each 📖 = 3 books

How many books on sports and travel does Lucas have?

A. 25

B. 21

C. 18

D. 15

14. Maria emptied her pockets and found some coins. She counted 48¢. Which of the following sets of coins did Maria have in her pocket?

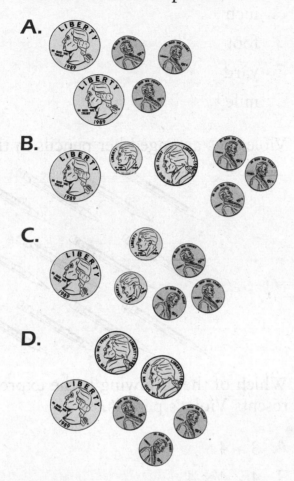

A.

B.

C.

D.

15. Mrs. Deyermond's class and Mrs. Cosgrove's class made paper chains to decorate the gym for a concert. The chain Mrs. Deyermond's class made had 398 links, and Mrs. Cosgrove's class made a chain with 484 links. How many more links did Mrs. Cosgrove's class make?

A. 196

B. 114

C. 96

D. 86

Go On

© Copyright 2012 by Barron's Educational Series, Inc.

16. Which unit of measure is the best to use to measure a new pencil?

   A. inch

   B. foot

   C. yard

   D. mile

17. Violet has arranged her pencils in the picture below.

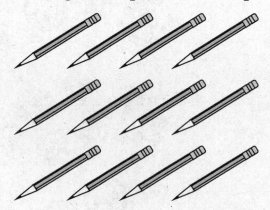

Which of the following is the expression that best represents Violet's pencils?

   A. 3 + 4

   B. 4 × 4 × 4

   C. 3 × 4

   D. 4 + 4

18. The clock in Charlie's classroom showed a quarter to 2. Which of the following is the clock in Charlie's room?

A.

B.

C.

D.

19. Which of the following is 800 + 5 in standard form?

A. 85

B. 805

C. 850

D. 855

20. Which of the following is the best estimate of the sum of 56 + 43 + 38?

A. 120

B. 137

C. 140

D. 150

Go On

© Copyright 2012 by Barron's Educational Series, Inc.

21. Which of the following squares is shaded exactly one-fourth?

A.

B.

C.

D.

22. Which of the following is the 3-dimensional name for these dice?

A. cube

B. cone

C. pyramid

D. cylinder

23. Which of the following will result in an even sum?

A. 4 + 3

B. 3 + 2

C. 2 + 5

D. 7 + 3

24. Peggy's room is 6 yards long. How many feet is that? Hint: 1 yard = 3 feet.

A. 6

B. 9

C. 18

D. 24

25. Which of the following is a division fact related to $2 \times 5 = 10$?

A. $5 \times 2 = 10$

B. $5 + 5 = 10$

C. $10 \div 2 = 5$

D. $10 + 2 = 12$

26. Which of the following figures could be a face of the drawing below?

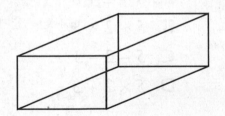

A.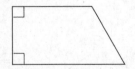

B.

C.

D.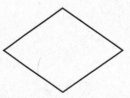

© Copyright 2012 by Barron's Educational Series, Inc.

27. How many lines of symmetry are there in the figure drawn below?

    A. 1

    B. 2

    C. 3

    D. 4

28. A bus was empty when 5 passengers got on at the first stop, 2 passengers got off at the second stop, and 3 passengers got on at the third stop.

    Which of the following is the best way to find out how many passengers were on the bus after the third stop?

    A. 5 + 2 + 3

    B. 5 − 2 + 3

    C. 5 − 2 − 3

    D. 5 × 2 + 3

29. Which of the following pair of figures are congruent?

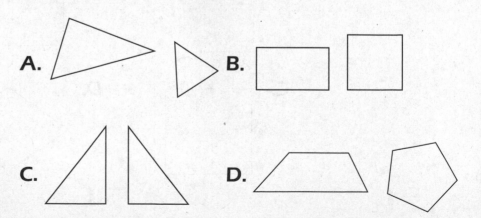

30. During a reading contest, Rainier read 598 pages, Andrew read 589 pages, Emmet read 597 pages, and Patrick read 588 pages. Which boy read the most pages in the contest?

A. Rainier

B. Andrew

C. Emmet

D. Patrick

31. Shannon laid out three number cards on the table. They looked like the ones shown below.

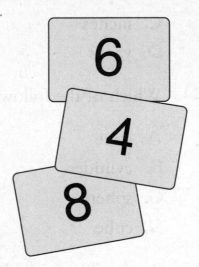

What is the largest number Shannon can make with an 8 in the tens place?

A. 864

B. 846

C. 486

D. 684

© Copyright 2012 by Barron's Educational Series, Inc.

Go On

32. Falon ordered a pizza for lunch. The pizza was cut into sixths. If she ate just one slice of pizza, what fraction of the pizza did Falon eat?

A. $\dfrac{1}{6}$ B. $\dfrac{1}{4}$

C. $\dfrac{1}{3}$ D. $\dfrac{1}{2}$

33. Which of the following is used to measure capacity?

A. pounds

B. cups

C. inches

D. yards

34. Which of the following has only square faces?

A. cone

B. cylinder

C. sphere

D. cube

35. Emelyn has 5 jelly beans in a bowl. One of them is red, and the rest of the jelly beans are pink. What fraction of Emelyn's jelly beans is red?

A. $\dfrac{1}{6}$ B. $\dfrac{1}{5}$

C. $\dfrac{1}{4}$ D. $\dfrac{1}{3}$

36. Kelly wrote the numbers shown below in her notebook.

$$6, \ 9, \ 12, \ 15, \ \underline{}, \ 21$$

What number goes in the blank to complete the pattern?

A. 16

B. 17

C. 18

D. 20

37. Which of the following is a true math statement?

A. $\frac{1}{2} < \frac{1}{4}$ **B.** $\frac{1}{8} > \frac{1}{6}$

C. $\frac{1}{4} < \frac{1}{6}$ **D.** $\frac{1}{2} > \frac{1}{3}$

38. Which set of coins below is worth 43¢?

A.

B.

C.

D.

© Copyright 2012 by Barron's Educational Series, Inc.

39. Katie loves to ride her horse. She feeds and cares for it every day. The weight of Katie's horse is most likely which of the following?

A. 100 ounces

B. 100 pounds

C. 1000 ounces

D. 1000 pounds

40. Isaiah goes to his piano lesson at the time shown on the clock below. At what time does Isaiah go to his lesson?

A. 3:40

B. 4:17

C. 3:20

D. 4:30

PART 2

Directions: Read each question in Part 2 carefully to be sure you answer it completely. If the question asks you to show your work, be sure to do so. You will receive partial credit for your work.

---

41. Daisy has a chocolate bar that she wants to share with her five friends. She would like everyone to receive $\frac{1}{6}$ of the candy. On the chocolate bar shown below, draw lines to show how Daisy's chocolate bar will look when she shares it.

42. Sahar has some coins in her pocket that have a value of $0.78. In the boxes below, draw coins to show two different groups of coins Sahar could have in her pocket.

43. The number line below has some markers with blanks below them. Place the fractions $\frac{1}{3}$, $\frac{1}{2}$, $\frac{1}{4}$, and $\frac{3}{4}$ in the blanks where they belong on the number line.

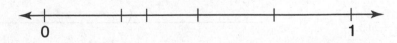

© Copyright 2012 by Barron's Educational Series, Inc.

44. Guess the number. Use the clues below to guess this number.

- My value is less than 400 and more than 300.
- The number in my tens place is one greater than the number in my ones place.
- The number in my ones place is 8.

On the lines below tell how you solved this problem.

_____

_____

_____

45. Study the figures drawn below. One of them has no lines of symmetry, and the other has two lines of symmetry.

A. Draw the lines of symmetry on the figure that has them.

B. On the lines below, explain why one of these figures has no lines of symmetry.

_____

_____

_____

_____

46. Chris has saved some shells she found on the beach. She has separated them by color. She has 18 white, 12 blue, 6 yellow, and 14 pink shells.

A. Complete the tally chart and frequency table below.

CHRIS'S BEACH SHELLS

Color	Tally	Number Collected
White		
Blue		
Yellow		
Pink		

B. Complete the bar graph below using the data in the frequency table. Be sure to label the axis.

Chris's Beach Shells

If you have time, you may review your work in this section only.

STOP

© Copyright 2012 by Barron's Educational Series, Inc.

ANSWERS

Part 1

1. **B.** 30 is an even number because there is a zero in the ones place.

2. **C.** A pentagon is a polygon with 5 sides.

3. **A.** The commutative property states that the order of the addends does not change the sum.

4. **C.** This is closest to the 3-inch mark

5. **A.** The identity property states that adding zero to a number always has a sum of that number.

6. **D.** Counting up from $3.95, a nickel makes $4 and another $1 makes $5, so $1.05 is the answer.

7. **D.** The hour hand is just past the 3, and the minute hand is on the 1.

8. **C.** One pet out of 4 is a rabbit, so the fraction is $\frac{1}{4}$.

9. **D.** $27 \div 9 = 3$

10. **A.** This is a cone, with one circular face and one vertex.

11. **C.** The 4 is in the tens place.

12. **D.** A bath tub holds a lot of water, and the biggest of these measures is the gallon.

13. **C.** There are 18 books because there are 6 book pictures and each of these is worth 3, so $6 \times 3 = 18$.

14. **C.** 25¢ + 10¢ + 10¢ + 3¢ = 48¢

15. **D.** $484 - 398 = 86$

16. **A.** The inch is the only one that is smaller than a pencil.

© Copyright 2012 by Barron's Educational Series, Inc.

17. C. This array shows 3 groups of 4 or 3 × 4.

18. D. A quarter to 2 is another way of saying 1:45. The minute hand is on the 9 and the hour hand is close to the 2.

19. B. 800 + 5 = 805

20. C. Using rounding to estimate, 56 becomes 60, 43 becomes 40, and 38 becomes 40 when rounded to the nearest 10. 60 + 40 + 40 = 140.

21. D. This is the only square evenly divided into 4 parts.

22. A. A cube has 6 square faces.

23. D. 7 + 3 = 10, which is an even number.

24. C. 3 × 6 = 18 feet.

25. C. 10 ÷ 2 = 5 is the only division fact to choose from.

26. B. The sides of this rectangular prism are all rectangles.

27. B. Every rectangle has only 2 lines of symmetry. They look like this.

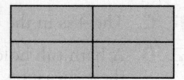

28. B. 5 passengers got on. When 2 got off, they were subtracted. Then 3 were added when 3 passengers got on.

29. C. The triangles are exactly the same.

30. A. 598 is greater than 589, 597, and 588.

31. D. 684 has an 8 in the tens place and the next largest digit is in the hundreds place. The smallest digit is in the ones place.

32. A. One piece out of 6 is $\frac{1}{6}$.

33. B. Cups measure how much liquid, which is capacity.

34. D. A cube has 6 square faces.

35. B. One out of 5 is $\frac{1}{5}$.

36. C. This pattern is adding 3 to each number.
$15 + 3 = 18$

37. D. The smaller the numerator is the larger the piece is, so $\frac{1}{2} > \frac{1}{3}$.

38. A. $25 + 10 + 5 + 3 = 43$

39. D. A horse is a big animal. 1,000 pounds is a reasonable weight.

40. C. The hour hand is just past the 3, and the minute hand is at 20 minutes.

Part 2

41.

42. The answer is any two combinations to make $0.78. The most common two might be 3 quarters and 3 pennies, and 2 quarters, 2 dimes, 1 nickel, and 3 pennies. Some students might even go for 78 pennies.

43.

$$\begin{array}{cccccc} 0 & \frac{1}{4} & \frac{1}{3} & \frac{1}{2} & \frac{3}{4} & 1 \end{array}$$

© Copyright 2012 by Barron's Educational Series, Inc.

44. The number is 398. Add 1 to 8 to get the tens place, and figure the hundreds place using the first clue. It must be 3 in order for the number to be greater than 300 but less than 400.

45.

The explanation should include the definition of a line of symmetry. It should also indicate that there is no way to draw a line on this figure and get two mirror images.

46. A.

CHRIS'S BEACH SHELLS

Color	Tally	Number Collected
White	IIII IIII IIII III	18
Blue	IIII IIII II	12
Yellow	IIII I	6
Pink	IIII IIII IIII	14

B.

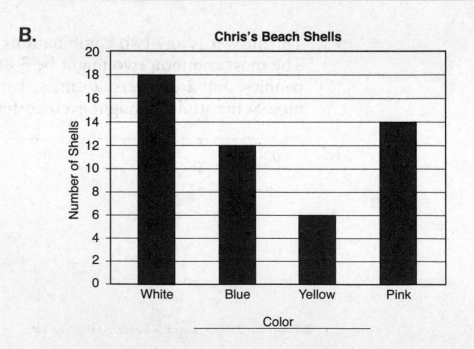

Chapter 13

Sample Test 2

ANSWER SHEET: SAMPLE TEST 2

PART I

1. Ⓐ Ⓑ Ⓒ Ⓓ	15. Ⓐ Ⓑ Ⓒ Ⓓ	29. Ⓐ Ⓑ Ⓒ Ⓓ
2. Ⓐ Ⓑ Ⓒ Ⓓ	16. Ⓐ Ⓑ Ⓒ Ⓓ	30. Ⓐ Ⓑ Ⓒ Ⓓ
3. Ⓐ Ⓑ Ⓒ Ⓓ	17. Ⓐ Ⓑ Ⓒ Ⓓ	31. Ⓐ Ⓑ Ⓒ Ⓓ
4. Ⓐ Ⓑ Ⓒ Ⓓ	18. Ⓐ Ⓑ Ⓒ Ⓓ	32. Ⓐ Ⓑ Ⓒ Ⓓ
5. Ⓐ Ⓑ Ⓒ Ⓓ	19. Ⓐ Ⓑ Ⓒ Ⓓ	33. Ⓐ Ⓑ Ⓒ Ⓓ
6. Ⓐ Ⓑ Ⓒ Ⓓ	20. Ⓐ Ⓑ Ⓒ Ⓓ	34. Ⓐ Ⓑ Ⓒ Ⓓ
7. Ⓐ Ⓑ Ⓒ Ⓓ	21. Ⓐ Ⓑ Ⓒ Ⓓ	35. Ⓐ Ⓑ Ⓒ Ⓓ
8. Ⓐ Ⓑ Ⓒ Ⓓ	22. Ⓐ Ⓑ Ⓒ Ⓓ	36. Ⓐ Ⓑ Ⓒ Ⓓ
9. Ⓐ Ⓑ Ⓒ Ⓓ	23. Ⓐ Ⓑ Ⓒ Ⓓ	37. Ⓐ Ⓑ Ⓒ Ⓓ
10. Ⓐ Ⓑ Ⓒ Ⓓ	24. Ⓐ Ⓑ Ⓒ Ⓓ	38. Ⓐ Ⓑ Ⓒ Ⓓ
11. Ⓐ Ⓑ Ⓒ Ⓓ	25. Ⓐ Ⓑ Ⓒ Ⓓ	39. Ⓐ Ⓑ Ⓒ Ⓓ
12. Ⓐ Ⓑ Ⓒ Ⓓ	26. Ⓐ Ⓑ Ⓒ Ⓓ	40. Ⓐ Ⓑ Ⓒ Ⓓ
13. Ⓐ Ⓑ Ⓒ Ⓓ	27. Ⓐ Ⓑ Ⓒ Ⓓ	
14. Ⓐ Ⓑ Ⓒ Ⓓ	28. Ⓐ Ⓑ Ⓒ Ⓓ	

PART II

Use the space in the test booklet to show your work and your answers.

© Copyright 2012 by Barron's Educational Series, Inc.

SAMPLE TEST 2

Answers on pages 225–230.

PART 1

Directions: Choose the best answer for each question and fill in the correct bubble on your answer sheet.

1. Which of the following is a rhombus?

A.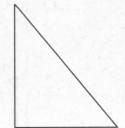

B.

C.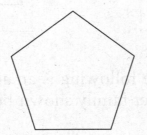

D.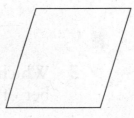

© Copyright 2012 by Barron's Educational Series, Inc.

GO ON

2. Diane lives on a street where all the houses have an even number. Which of the following could be Diane's front door?

A.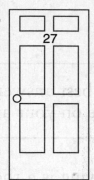
27

B.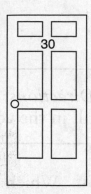
30

C.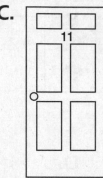
11

D.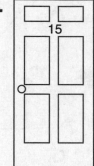
15

3. Which of the following is an **addition** sentence member of the fact family shown below?

$$9 - 3 = 6$$

A. $9 - 5 = 4$

B. $9 + 3 = 12$

C. $3 + 6 = 9$

D. $9 + 6 = 15$

4. Which of the objects below would best be measured in inches?

A.

B.

C.

D.

5. Which of the following is 502 written in expanded form?

A. 50 + 2

B. 5 + 0 + 2

C. 500 + 2

D. 500 + 20

6. Which of the following clocks shows 10:30?

A.

B.

C.

D.

GO ON

© Copyright 2012 by Barron's Educational Series, Inc.

7. The picture below shows a can of soup on the shelf in Blake's kitchen.

Which of the following could be the shape on the bottom of the can.

A.

B.

C.
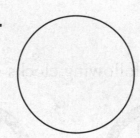

D.
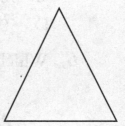

8. Which of the following is an example of the commutative property of addition?

A. 8 + 9 = 9 + 8

B. 6 − 0 = 6

C. 7 + 1 = 8

D. (2 + 3) + 4 = 2 + (3 + 4)

9. Which of the following is a set of only odd numbers?

    A. 9, 8, 7, 6

    B. 28, 30, 32, 34

    C. 16, 17, 18, 19

    D. 11, 13, 15, 17

10. Which multiplication problem is shown in the array below?

    A. 3 + 4

    B. 3 × 4

    C. 2 × 3

    D. 4 + 4 + 4

11. Nick was given the object pictured below for his birthday.

Which is the three-dimensional name for this object?

    A. cone

    B. cylinder

    C. prism

    D. sphere

GO ON

© Copyright 2012 by Barron's Educational Series, Inc.

12. Which of the following rectangles is divided into thirds?

A.

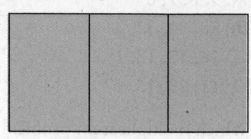

B.

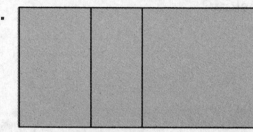

C.

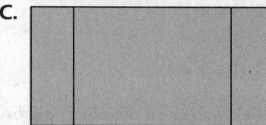

D.

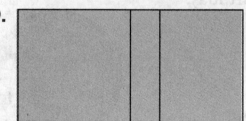

13. Which of the following fractions has the greatest value?

A. $\frac{1}{5}$

B. $\frac{1}{10}$

C. $\frac{1}{3}$

D. $\frac{1}{2}$

14. Which of the following figures has two lines of symmetry?

A.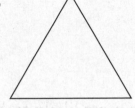

B.

C.

D.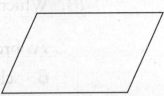

15. Which number completes the following expression?

$$738 > \underline{\hspace{1cm}}$$

A. 742

B. 761

C. 729

D. 740

16. Which of the following tools would you use to measure capacity?

A. scale

B. cup

C. pound

D. mile

GO ON

© Copyright 2012 by Barron's Educational Series, Inc.

17. What is the value of the 3 in 539?

    A. 0

    B. 3

    C. 30

    D. 300

18. Which event is most likely to happen at 6:00 P.M.?

    A. breakfast

    B. school recess

    C. math class

    D. dinner

19. Which number can be placed on the star on the number line below?

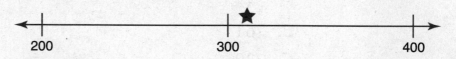

    A. 250

    B. 300

    C. 310

    D. 350

20. Karen is baking cookies and has placed 4 rows of cookies on her baking sheet. If each row has 6 cookies, how many cookies is Karen baking?

    A. 4

    B. 6

    C. 10

    D. 24

21. Kurt has collected shells from the beach. He has placed them in the bags below. How many shells did Kurt collect in all?

Shells 165   Shells 345

A. 400

B. 410

C. 500

D. 510

22. Josie bought a 2-pound bag of potatoes for baking. If there are 16 ounces in one pound, how many ounces did she buy?

A. 16

B. 18

C. 32

D. 33

23. Sue owns four pets. She has two bunnies, a cat, and a dog. Which of the following is a fraction name for the number of bunnies Sue has?

A. $\frac{1}{2}$

B. $\frac{1}{3}$

C. $\frac{1}{4}$

D. $\frac{1}{5}$

GO ON

© Copyright 2012 by Barron's Educational Series, Inc.

24. Fran was skip counting on her way home from school. This is what she said, "20, 22, 24, 28, 30, 32, 34, ..."

What number did Fran leave out?

A. 21

B. 25

C. 26

D. 29

25. Jeff was counting the change he found in his pocket. He had four coins that added up to 26¢. Which of the following sets of coins did Jeff have in his pocket?

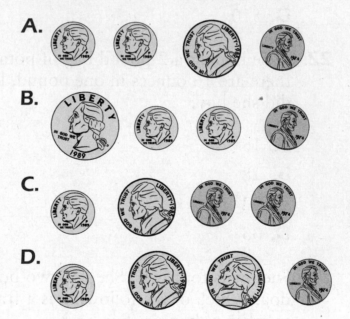

A.

B.

C.

D.

26. Which of the following statements is true about the number sentence shown below?

$$8 \times 0 =$$

A. The product is 8.

B. The product is 0.

C. The product is greater than 8.

D. The product is 80.

27. Use your ruler to help you with this question.

How long is this paintbrush to the nearest $\frac{1}{2}$ inch?

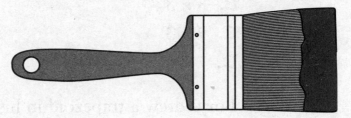

A. $2\frac{1}{2}$ inches

B. 3 inches

C. $3\frac{1}{2}$ inches

D. 4 inches

28. Gavin wrote the number pattern below:

50, 75, 125, 150, 175, 200, 225

What number did Gavin leave out?

A. 215

B. 180

C. 135

D. 100

29. Ella noticed that she is taller than $\frac{3}{4}$ of her friends. Using that fraction, which of the following is true?

A. Ella has 7 friends.

B. Ella is the most popular girl in her class.

C. Ella is taller than 3 of her friends.

D. Ella is taller than 4 of her friends.

GO ON

© Copyright 2012 by Barron's Educational Series, Inc.

30. Which of the following has the greatest product?

A. 6 × 5

B. 8 × 3

C. 9 × 3

D. 7 × 4

31. Borys drew a trapezoid in his notebook. Which of the following could be Borys's drawing?

A.
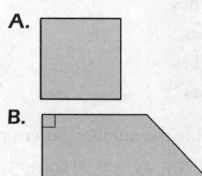

B.
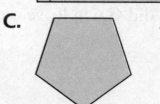

C.
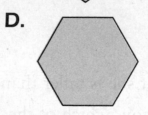

D.

32. Leyla loves to visit the Bengal tiger exhibit at the zoo. She estimates that her favorite tiger weighs which of the following?

A. 60 ounces

B. 60 pounds

C. 600 ounces

D. 600 pounds

33. The spring play was very popular this year. There were eight hundred nine tickets sold. Which of the following is the number of tickets sold written in standard form?

A. 89

B. 98

C. 809

D. 890

34. Sarah has started a collection of coins. She has more than 546 coins. Which of the following could be the number of coins Sarah has collected?

A. 501

B. 564

C. 543

D. 539

35. Paul has six cookies with eight chocolate chips in each cookie. Which number sentence should he use to find out how many chocolate chips he has altogether?

A. 6×8

B. $6 + 8$

C. $8 \div 6$

D. $8 - 6$

36. Which of the following number sentences will result in an even sum or difference?

A. $8 + 3$

B. $9 - 4$

C. $11 + 3$

D. $5 - 2$

GO ON

© Copyright 2012 by Barron's Educational Series, Inc.

37. Brenda got out a stick of butter for her toast.

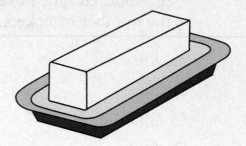

Which of the following shapes are faces of this stick?

A.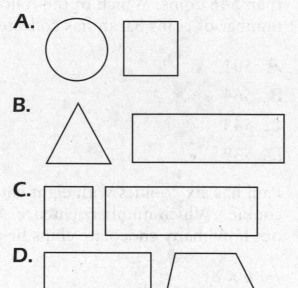

B.

C.

D.

38. Which of the following is the best way to measure the capacity of a swimming pool?

A. pint

B. quart

C. cup

D. gallon

39. Jan has a giant cookie she wants to share with three friends. Which drawing below shows Jan's cookie if she divides it into fourths?

A.

B.

C.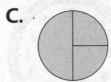

D.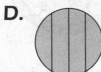

40. Maria wants to get up at a quarter after seven. Which digital clock shows that time?

A.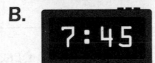
7:25

B.
7:45

C.
7:15

D.
7:30

© Copyright 2012 by Barron's Educational Series, Inc.

PART 2

Directions: Read each question in Part 2 carefully to be sure you answer it completely. If the question asks you to show your work, be sure to do so. You will receive partial credit for your work.

41. Draw hands on the clock face below to show the time half past 5.

If this were a digital clock, what time would it say?

_____

On the lines below explain why you drew the hands where you did to show that time.

_____

_____

42. Study the pattern of numbers below.

55, 52, 49, ___, 43, 40, ___, 34

What two numbers belong in the pattern?

_____, _____

On the lines below explain this pattern and how you got your answer.

_____

_____

_____

43. The Central Zoo gave guided tours today. There were 38 people for the first tour, 24 people for the second tour, 32 people for the third tour, and 27 people for the last tour. About how many people went on guided tours at the zoo today? Show your work.

Answer: **About**_____ people

GO ON

© Copyright 2012 by Barron's Educational Series, Inc.

44. A. On the grid below draw 2 congruent squares.

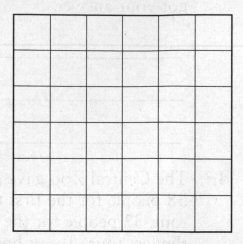

B. On the grid below draw 2 similar but not congruent squares.

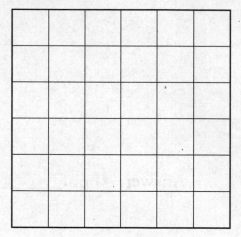

45. Jessie is selling colored pencils at the school store. She has noticed that some colors are more popular than others and has made the tally chart below.

COLORED PENCILS SOLD AT THE SCHOOL STORE

Color	Tally	Total
Red	ⅢⅢ I	
Orange	IIII	
Yellow	ⅢⅢ III	
Green	II	
Blue	ⅢⅢ ⅢⅢ II	
Purple	ⅢⅢ III	

A. Complete the chart by entering the totals for each color.

B. Use the grid below to make your own pictograph of the number of colored pencils sold.

COLORED PENCILS SOLD AT THE SCHOOL STORE

Color	Number of Pencils
Red	
Orange	
Yellow	
Green	
Blue	
Purple	

Key: _____ = _____ pencils

GO ON

© Copyright 2012 by Barron's Educational Series, Inc.

46. The graph below shows the amount of money made at the school store in one week.

MONEY MADE AT THE SCHOOL STORE

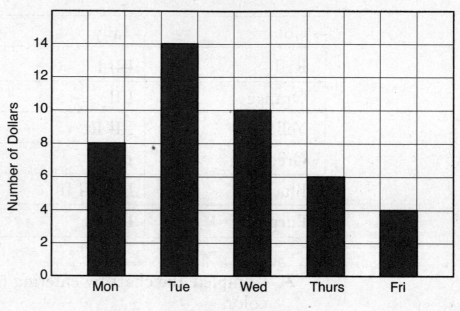

A. How much more money did the store make on Tuesday than on Monday? _____

B. On which 2 days did the store make the same amount together as it did on Tuesday? Show your work in the space below.

Answer: _____

If you have time, you may review your work in this section only.

ANSWERS

Part 1

1. D. A rhombus has 4 equal sides.

2. B. 30 has a 0 in the ones place, so it is even.

3. C. This is an addition sentence that includes 3, 6, and 9.

4. A. A pencil is the only object small enough for inches.

5. C. $500 + 2 = 502$

6. A. The hour hand is halfway past the 10, and the minute hand is on the 6.

7. C. The can is a cylinder, so the base is a circle.

8. A. The commutative property states that the order of the addends does not change the sum.

9. D. There is a 1, 3, 5, and 7 in the ones place of each of these numbers. They are all odd.

10. B. 3×4 can be read 3 groups of 4 which is what this array shows.

11. D. This is a globe, which is a sphere. Any object shaped like a ball is a sphere.

12. A. This is the only rectangle divided into equal parts.

13. D. $\frac{1}{2}$ has the greatest value because only 2 halves make a whole. The greater the denominator, the smaller the pieces.

14. C. A rectangle has only 2 lines of symmetry, one running vertically down the center and one running horizontally through the middle.

15. C. 729 is the only choice that is less than 738.

© Copyright 2012 by Barron's Educational Series, Inc.

16. B. Capacity is liquid measure, so the cup is the only answer.

17. C. 3 is in the tens place, and 3 tens is 30.

18. D. P.M. means after noon and 6:00 P.M. is evening time, perfect for eating dinner.

19. C. 310 is greater than 300 but less than 350 so it is just to the right of 300.

20. D. If there are 4 rows of 6 cookies in each row, then $4 \times 6 = 24$.

21. D. The sum of 165 and 345 is 510.

22. C. The bag weighs 2 pounds, so $16 + 16 = 32$.

23. A. Two of four animals are bunnies, so $\frac{1}{2}$ of her animals are bunnies.

24. C. Fran is skip counting by 2, so she left out 26.

25. A. 2 dimes + 1 nickel + 1 penny = 26¢

26. B. This is the zero property. Anything multiplied by zero is zero.

27. C. The paintbrush is just a little less than $3\frac{1}{2}$ inches.

28. D. This pattern is counting by 25. 50, 75, 100, 125, 150, 175, 200, 225

29. C. The fraction $\frac{3}{4}$ means 3 out of 4, so Ella is taller than 3 of her 4 friends.

30. A. $6 \times 5 = 30$, $8 \times 3 = 24$, $9 \times 3 = 27$, $7 \times 4 = 28$

31. B. The top and bottom sides of this trapezoid are parallel.

32. D. A tiger weighs about 600 pounds.

33. C. Eight hundred nine = 809

34. B. $564 > 546$

35. A. Multiplication is the correct operation to use.

36. C. 11 + 3 = 14

37. C. This stick of butter has faces in the shape of rectangles and squares.

38. D. This is the largest liquid measurement, and the pool is a big thing to measure.

39. A. Jan will divide her cookie into four equal parts.

40. C. A quarter hour is 15 minutes, so the time is 7:15.

Part 2

41.

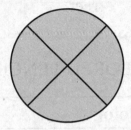

The time showing is 5:30.

Explanation: The hour hand must be halfway between the 5 and the 6 to show that the hour is half gone, and the minute hand must be at the 30 showing that 30 minutes have passed.

42. The two numbers that belong in the pattern are 46 and 37.

Explanation: This pattern is counting back by 3 each time, so 3 less than 49 is 46 and 3 less than 40 is 37.

© Copyright 2012 by Barron's Educational Series, Inc.

43. To find out about how many people took tours you must round each number to the nearest ten first. 38 becomes 40, 24 becomes 20, 32 becomes 30, and 27 becomes 30. The sum of 40 + 20 + 30 + 30 = 120.

The answer is about 120 people.

44. A. Any 2 squares that are the same size are correct.

B. Any 2 squares that are *not* the same size are correct.

45. A.

COLORED PENCILS SOLD AT THE SCHOOL STORE

Color	Tally	Total
Red	IIII I	6
Orange	IIII	4
Yellow	IIII III	8
Green	II	2
Blue	IIII IIII II	12
Purple	IIII III	8

B.

COLORED PENCILS SOLD AT THE SCHOOL STORE

Color	Number of Pencils
Red	
Orange	
Yellow	
Green	
Blue	
Purple	

Key - = 2 pencils

The student could also use one picture for each pencil sold. The pictures must all be the same, however.

© Copyright 2012 by Barron's Educational Series, Inc.

46.

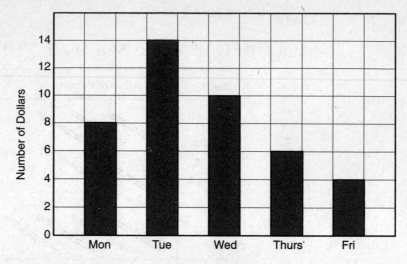

A. $14 − $8 = $6

B. On Wednesday $10 was made, and on Friday $4 was made. $10 + $4 = $14

Answer: Wednesday and Friday

Index